Sustainability Creates New Prosperity

Karin Feiler (ed.)

*European Forum on Sustainability
of the Club of Rome*

Sustainability Creates New Prosperity

Basis for a New World Order, New Economics
and Environmental Protection

*Review by Members of The Club of Rome
and International Experts*

Preface by Klaus Töpfer

With Contributions by
Martin Bartenstein, Orio Giarini, Hans Küng, Uwe Möller,
Patrick M. Liedtke, Mahendra Shah, Franz Josef Radermacher,
Walter R. Stahel, Ernst Ulrich von Weizsäcker

PETER LANG

Frankfurt am Main · Berlin · Bern · Bruxelles · New York · Oxford · Wien

Bibliographic Information published by Die Deutsche Bibliothek
Die Deutsche Bibliothek lists this publication in the Deutsche
Nationalbibliografie; detailed bibliographic data is available in
the internet at <http://dnb.ddb.de>.

The edition of this Review by Members of
The Club of Rome
was supported by the
United Nations Information Service (UNIS), Vienna.

ISBN 3-631-51973-7
US-ISBN 0-8204-6549-6

© Peter Lang GmbH
Europäischer Verlag der Wissenschaften
Frankfurt am Main 2004
All rights reserved.

Printed in Germany 1 2 4 5 6 7

www.peterlang.de

Preface

In 1972 the Report of The Club of Rome "The Limits to Growth" conveyed with its provoking scenarios the clear "message" that humankind is facing finite natural resources. Since then we know that we have to leave our traditional resource-consuming path of production and life-style. Prosperity for all and for coming generations will only be viable in an socio-economic environment leading to sustainable development.

Developed on the initiative of the European Forum for Sustainability of The Club of Rome – a joint venture of the European Support Center of The Club of Rome and the Austrian Federal Ministry of Economy and Labor –, this book wants to contribute to the ongoing discussion how we can build the framework for an eco-social market economy indispensable for sustainable development. This constitutes a great challenge to all of us be it as citizen, voter or politician, be it as consumer, worker or entrepreneur. We all have to change our way of thinking and our attitudes.

This book wants to stimulate us to do so and to take responsibility! We appreciate very much the initiative of Karin Feiler who succeeded in bringing together leading experts in the field of sustainability, among others several members of The Club of Rome. Hopefully this book with its insiring ideas will find an interested readership.

Uwe Möller
Secretary General, The Club of Rome

Authors

Dr. Martin Bartenstein, Federal Minister for Economy and Labour in Austria

Dr. Karin Feiler, Head of Department for Economic Sustainable Development, BMWA, Vienna

Professor Orio Giarini, Member of the Club of Rome in Genf/Triest

Dipl.-Volksw. Jochen Jagob, Assistent at the Institute for Economics of TU Darmstadt

Dr. Petra Gruber, Head of the Institute for Environment, Development and Peace, Vienna

Professor Dr. Hans Küng, Head of the Institute for World Ethos, University Tübingen

Patrick M. Liedtke, Vice-President of the European Support Centre of the Club of Rome, Geneva

Uwe Möller, General Secretary of the Club of Rome, Hamburg

Professor Dr. Dr. Franz Josef Radermacher, Head of the Forschungsinstituts für anwendungsorientierte Wissensverarbeitung in Ulm, Member of the Club of Rome

Former Vice-Chancellor. Dipl.-Ing. Dr. h. c. Josef Riegler, President of the Ecosocial Forum of Austria in Vienna/Graz

Professor Dr. Dr. h. c. Bert Rürup, Chair for Fiscal and Economic Policy at the Institute for Economics (Institut für Volkswirtschaftslehre of TU Darmstadt)

Professor Dr. Josef Schmid, Science for Demographic Development, Otto-Friedrich-University in Bamberg

Mahendra Shah, Ph.D., International Institute for Applied Systems Analysis (IIASA), Laxenburg;

Walter R. Stahel, Head of the Risk Management Programme of the Geneva Association, Founder Director of the Product-Life Institute, Geneva

Professor Dr. Klaus Töpfer, Executive Director, United Nations Environment Programme, Nairobi

Professor Dr. Ernst Ulrich von Weizsäcker, Chairman Bundestag Environment Committee, Germany, Member of the Club of Rome

Michael Windfuhr, FIAN Executive Director der Internationalen Menschenrechtsorganisation für das Recht sich zu ernähren FIAN-International und Maartje van Galen, FIAN-International

Mag. Gabriele Zöbl, Head of the European Support Centre of the Club of Rome, Vienna

Contents

V. Economic and Socio-political Responsibility

VI. Social and Human Responsibility

VII. Closing Word

I. Preface

The Significance of Sustainability – Sustainability of Prosperity

Klaus Töpfer

The Environment as a Contributor to Sustainable Prosperity!

2002, the organization United Nations Environmental Programme in Nairobi published the Global Environment Outlook-3. It highlights that key environmental indicators continue to deteriorate in many parts of the world – particularly in the developing world. The millions who die from water pollution related illnesses and the lost production from degraded lands are testimony to this, and as population increases in some areas and environmental degradation worsens, the situation is deteriorating for millions of the poorest people on our planet, our home.

Sustainable prosperity depends on there being sufficient resources to generate a sustainable stream of benefits to humanity, over time and location. Most people appreciate the need for sufficient resources, but more need to appreciate that the environment is a key pillar of sustainable prosperity, rather than a competitor. While a response to criticisms of unsustainable use of the environment may be that there is no other choice but to use the environment in an unsustainable way, this view of the environment as a competitor misses the critical point that the environment is fundamental to sustainable development. As the World Bank put it in relation to water: "the environment [is] not just another consumptive user of water but the water resource itself and ... degrading the quantity and quality of water in rivers, lakes, wetlands and aquifers can inextricably alter the water resources system and its associated biota, affecting present and future generations." In addition, when some say there is no other choice but to degrade the environment, they almost always say this without calculating the costs of environmentally unsustainable resources use. Worse, it is usually the poor who suffer the most from environmental degradation – with UNEP's work on poverty and the environment highlighting the link between unsustainable resource use and poverty.

Using an example from the land and water sectors, perhaps the best-known example of the economic costs of environmentally unsustainable resource use is the case of the Aral Sea basin. There, the environmentally unsustainable use of freshwater resources for agricultural irrigation decreased lake water levels and quality to such an extent that the previously robust fishing industry completely collapsed. Nearly all investment in this industry now lies idle. In addition, the inefficient use of irrigation water in this semi-arid region has lead to land salinisation and a subsequent large decrease in agricultural production.

In a nutshell, we need to move beyond the concept of environmental protection to sustainable resource management, in order to achieve sustainable development and sustainable prosperity.

And to achieve sustainable resource use requires integrated environmental, economic and social approaches. Thus, policy, law and management in the key resource using sectors is the priority focus – the task cannot be left to the environment ministries.

Moving from the key sectors to the broader economic framework, how does the concept of sustainable prosperity relate to classic economic theory? The main thrusts of economic policy today are based on economic theories developed before the concept of sustainable development existed. Thus, while economics has been under-utilised at a microeconomic level in *assisting* us to achieve sustainable development, conventional economic theory and practice are not, I believe, suitable as the *determinant* of how we manage our environmental resources. That is, 'leave it to the markets' is not an appropriate response to the issue of how we achieve sustainable prosperity. While various useful methods of incorporating non-market valuations and pricing environmental resources have been developed, the more fundamental issues of the ability of the macro-level application of conventional economic models to achieve sustainable development and prosperity have not been resolved. Nor are they likely to be. Having said that, I would like to stress that greater use of economics at a micro-economic level would greatly assist in improving the sustainability of resource use. For example, using pricing to encourage more efficient use of land and water and less pollution.

The issue of inter- and intra-generational equity is also poorly addressed in conventional economic models – particularly in relation to sustainable development in developing countries.

So in summary, we need to make more use of economics at the micro-economic level to achieve sustainable prosperity, while developing more appropriate economic models at the macro-level that explicitly incorporate the concept of sustainable development and better cater for inter- and intra-generational issues.

Developing countries are understandably wary about the application of economic techniques and models developed in the West – not just due to their concerns about the appropriateness of particular techniques and models, but also because of their uneven application. For example, Western pressure on developing countries to reduce trade tariffs, subsidies and increase prices for water and electricity are in some cases grossly hypocritical, given for example, agricultural,

14

water and fishing subsidies in the EU and US. Thus, equity through consistency in the application of economics and globalization is also necessary.

While some look to the transformation from industrial based societies to service based societies to achieve sustainable prosperity, we will always need the 'old' industries of, for example, metal and chemicals, to produce many necessities. We can become more efficient in the production of items, and the use of energy per unit of output, and we must do so. But fundamental questions regarding production and consumption remain – what are the implications of the entire planet achieving Western levels of consumption patterns? In general, the implications for the environment and our resource base are disturbing. It is completely unrealistic and grotesquely unfair to imply that developing countries should not seek Western standards of living. Thus it is imperative that the West – which established the pattern of production and consumption that potentially threatens the entire planet – puts far more effort in developing a modified pattern that will help all countries achieve an equitable and sustainable standard of living. This is vital to achieving sustainable prosperity and peace. For, make no mistake about this: There is a real potential for serious conflict between the haves and have-nots unless we achieve an equitable and sustainable prosperity for all.

II. Introduction

Sustainability: Expectations and Reality

Uwe Möller

"Limits to Growth" – Something to Think About

The CLUB OF ROME – founded in 1968 – with its report "The Limits to Growth" published in 1972 has contributed decisively to the initiation of a new thought dimension that is today related to the expressions 'sustainability' or 'sustainable development'.

At the time of its publication, this report to the CLUB OF ROME posed a completely new and revolutionary question regarding the finiteness of natural resources. "The Limits to Growth" was written in a period when economic growth and its ability to create more wealth for all was taken for granted.

The following issues drove the founders of the CLUB OF ROME: are natural resources and energy sufficiently available to maintain economic growth? Is the natural environment – air, water, soil and the variety of species – able to deal with an increased level of 'stress factors', as a result of economic activity?

The report "The Limits to Growth"enjoyed great popularity because it was based on a "mathematical world model" which was able to generate quantitative predictions of future scenarios far into the 21st century. Twelve million copies were published in more than 25 languages worldwide.

The computer explored a new realm for economics and the social sciences, as until then, social and economic reality could only be described and explained in simple verbal models. From then on, an all-encompassing, complex system could be implemented. Calculating with world models was something new and it was also new within the question area of "The Limits to Growth". Therefore, the scientific world had to "check on" the calculations.

This resulted in strong criticism for the world model and the future scenarios presented in the report. The parameters and the assumed relationships between the variables were regarded as inaccurate, including the idea in the first place! If the "Limits to Growth" had been presented only verbally, without the world model, the paper would definitely not have received the same attention and publicity.

Even though the paper "Limits to Growth" was criticised initially in the academic world, it offered decisive inputs to create a strong awareness of the global relationship between economics and ecology. Today, it is accepted that the resources on this planet are finite. We are aware that we have to use resources in a sustainable fashion if we do not want to destroy the fundamental basis of survival for future generations.

In 1972, "Limits to Growth" also made predictions about the decades preceding the millennium. The authors, Dennis and Donella Meadows, published their new book "The New Limits to Growth" in 1992 in which they drew conclusions using their calculations and predictions of 1972 in relation to the data available up to 1990. The first report estimated that with a growing world population and an increasing production of goods and services, shortages in raw materials and energy were soon to appear. However, in 1992, the new book showed that these worries were currently not as serious as expected in 1972. This was firstly because the amount of resources available had been underestimated. Secondly, technological advances had resulted in a more efficient use of resources. And also as recycling practices had become more and more popular, the amount of new resources required could be reduced. However, fossil resources still account for up to 90 percent of our energy consumption. In the exploitation of natural oil and natural gas the currently tapped, easily accessible and cheap sources will run out comparatively soon. Nonetheless, this fact has still not led to a search for more sensible ways of energy use. At present, it is doubtful whether new technologies for using energy will be developed in time to resolve threatening upcoming shortages.

The Destruction of Nature's Capital

The data in 1990 shows alarmingly that the strain on and the destruction of the natural environment in the previous 20 years had increased twice as fast as was feared in 1972. The "consumption of nature" results from the "economies of waste" of the rich. The population of 1,5 billion in the "north" accounts for 80 percent of economic production along with the resulting use of natural resources while the poor "south" with its 4,5 billion inhabitants accounts for less than one fifth of economic production. The excessive use of resources of the "economies of survival of the poor" leads to:

– the destruction of fertile soil, which threatens the potential for the production of the biomass which is essential for the survival of mankind,
– the destruction of essential water reserves, which increases the possibility of conflicts in the developing world,
– the pollution and over-fishing of rivers and oceans,

– the destruction of genetic potential due to the accelerated diminishing of the variety of species, and

– the increased risk of climate changes.

The "north" does not recognise these threats that easily, as they mainly endanger the survival of the population of the southern hemisphere, which experiences hunger, under-nourishment, disease and social misery as a result. This leads to migration and distribution fights, violence, terror and military conflict that destroy additional human, social and natural capital. The materially well off and politically stable "north", hence also ourselves in Europe, still recognise these issues mainly only as "distant problems". This, despite migration patterns of asylum seekers entering Europe as well as the increasing dangers resulting from terrorism threats increasingly prove that the "one world" or the "global village" where we all live is not "political lyrics" but "firm reality".

Poverty – Justice and Sustainability

Over the past years, poverty has further increased in many regions of the developing world and poses a substantial threat to peace and stability in the world. Even in highly developed societies in the "north", the inequality of the distribution of income and wealth creates a social time bomb. Poverty in the "south", which results mainly from economic underdevelopment, has a variety of causes: in many cases, ruling groups in power are not interested in economic development as it inevitably leads to the rise of a middle class in civil society which would question the authoritarian power structure. These "elites" apply their philosophy of power primarily in order to gain positions and power and not to further socio-economic development. Therefore, most of their energy is needed for power battles and not concentrated in the socio-economic development.

But the "north" also complicates the economic development of the "south". The selfish protectionism presents a substantial obstacle as it hinders the integration of upcoming economies in the developing world into the growing world markets. Additionally, current development aid has largely been implemented contraproductively. It has often been invested into less sensible large-scale projects favouring certain power groups.

Meanwhile, we know what has to be done on the "poverty front". The "north" has to make it easier for the upcoming economies of the "south" to participate in the international division of labour and the world markets by removing obstacles and providing special support. "Trade for aid" should still be the priority in development politics, while development aid essentially serves to support economic activities of small and medium sized enterprises, especially in rural areas. It should be urged that, especially in metropolitan regions where most economic activities take place in an "informal" sector, it is essential to create a legal framework for potential future development. Apart from education, initiatives that aim to build a civil society and improve the role of women should be supported. The latter not only helps to lower birth rates but also to strengthen the socio-economic position of women. This could produce decisive impulses for economic development, as in many regions of the developing world, women, who often work in harsh conditions, account for up to two thirds of economic production.

No Sustainability without Peace

In recent years, the global political scene has been characterised by a rising number of power struggles and violent conflicts. These have numerous origins: tribal and clan differences, nationalistic as well as religious-cultural hostilities and growing socio-economic disparities. They are accompanied by the breakdown of legal state power and authority. Criminal and terrorist power structures increasingly form and conduct private warfare aided by arms proliferation.

What has to be done? – First of all, an improved crisis management is needed that is also able to create and protect peace and if necessary, by military means. The US is the decisive power that is able and willing to intervene in conflicts by military means. However, the US – torn between unilateralism and multilaterism – also requires partners as the (still incomplete) United Nations framework. Europe too will increasingly have to assume global political responsibility when it comes to security issues. This can only be achieved through unified European Union security and foreign policies, which can draw on integrated military potential.

What has to be done: we need better crisis management within the UN framework as well as effective local security systems. The proliferation of arms has to be stopped. International law and human rights norms have to be further enforced. Education favouring peace and tolerance will have additional positive effects and non-governmental organisations could play a stronger role in the development of a global civil society.

Unsatisfying "Outcome of Sustainability"

The "outcome of sustainability" is unsatisfying. Sustainable development has widely been accepted as the most important goal in securing the survival of future generations. Political parties, groups in society, industry, government and international organisations have all declared sustainable development a priority.

Steps towards global sustainable development and the resulting duties and obligations were agreed upon at many international conferences, especially in 1992 at the World Summit in Rio, but implementation is sadly lacking. The World Summit in Johannesburg in September 2002, which drew a conclusion on the efforts that had been made towards sustainable development since Rio, showed this blatantly. In Johannesburg, future goals were defined vaguely with little obligations.

Whether it will be possible to ensure sustainable development for humankind in the "global village" will be decided in the growing markets of the developing world. At present, the still poor and growing population understandably strives for higher living standards, which the affluent societies of the "north" enjoy. However, we know that the strain that such a development would put on the environment, the raw material and energy reserves, would be unsustainable. The current consumption of the environment, of which the "north" accounts for up to 80 percent, is already destroying the natural living conditions for future generations. Our style of life is not transferable to the "south". The "Asian tiger" states, which have followed our industrialisation pattern with its high growth rates successfully, are often envied but cannot serve as examples. Their development yielded high resource consumption and a substantial destruction of the environment. This is especially visible in the case of China. Some Chinese regions are achieving "successful" economic development with a high burden on the environment. That is due to the fact that the energy supplied to serve the demand is mainly generated through the use of dirty coal technologies. The Chinese government has begun to realise the severe problems (e.g. health threatening air pollution) the population is experiencing as a result.

Efficiency Revolution in the Use of Resources

If the growing markets in the developing world play the decisive role regarding the question of resources, then transnational corporations hold a strategic position. They, as global players, have a global network and the necessary capacities for technology transfer. In the meantime, these companies have to fuel the efficiency revolution in the use of resources by creating a totally changed and resource-efficient collection of goods and services, which are marketable and exportable.

Those who want to sell in the world markets in the future know that this is only possible with products and processes that fulfil resource and environmental requirements. If an ecological tax reform makes resources more expensive, then the technologies mentioned will be implemented more quickly. The current capital and technology intensive production of goods with its low costs on raw materials, energy and the environment and with its wasteful use of resources is doomed to fail. Technologies with closed resource and life circles require labour intensive processes. The standard of living can no longer be achieved through overexploitation of natural resources, but has to be reached through "work alone". More environment means more work.

Energy plays a central role in the dematerialisation process. It has a strong influence on the movement of goods: the cheaper and greater its supply, the more it encourages movement and transport of goods and services. It becomes clear where dematerialisation has to start when we look at the current comforts civilisation enjoys: living comfort and mobility alone account for roughly two-thirds of energy and resource consumption.

We are still far from accepting the necessary ecological change and its consequences. Recognising this necessary change is now an obvious part of the fundamental program of every political party, but daily political life is different, as sustainability has yet to make a substantial impact on voting preferences. It is possible to talk openly and sympathetically to the enlightened and educated groups of society about the necessity of sustainable development. These groups always state that they are aware of the responsibility they have towards future generations. But would they change their consumption habits in favour of "green products and services"? With their higher standards of living, they possess substantial purchasing power and would be able to accelerate the restructuring of the economy towards sustainability and higher resource and energy efficiency. Companies are already anticipating the demand for sustainable goods and services; they will be able to offer "green" products in many markets.

However, the short-term oriented shareholder value approach that originates from the capital markets makes it harder for CEOs and managers to plan forward-looking long-term investments in sustainable products. Roughly the same applies as in consumer markets: in the same way the consumer has to be convinced of the advantages of sustainable demand, it has to be made clear to the shareholder that long-term profits can increasingly only be secured by sustainable investments.

Change in Life-Styles

The dematerialisation of our economy and our standards of living contains two different aspects. Firstly, technological advances can improve the efficiency of resources. Secondly, it also requires a change in life style that entails material moderation and places greater emphasis on immaterial values. Why, for example, does society place such a high emotional value on cars exceeding their real, rational usefulness in terms of mobility? This has a substantial effect on the use of resources and the environment, while culture and education lack the necessary funding. Could individual and social models not be characterised stronger in terms of "being" than in terms of "owning"? Will the limits of material growths not force us anyway? And would this not in the end create a "better and more liveable world"?

We should not forget that we in the "north" hold a responsible model function for the upcoming "south" and we have to develop a new model for living as a convincing mixture of efficiency and sufficiency and also live it.

Sustainable Development:
A Contradiction in Terms or an Economic Necessity

Ernst Ulrich von Weizsäcker

Sustainable development, in the long run, is by definition an economic necessity. Its essential meaning is that we should not destroy the natural basis of our wealth so that development can go on forever. But if development is understood as the present type of economic growth, "sustainable development" is a contradiction in terms.

The Club of Rome with its famous *Limits to Growth*[1] Report pointed out the problem without using the term sustainable development. The report showed that economic growth could not go on forever. The answer to *Limits to Growth* was a major paradigm shift in the application of technology after the publication in the year 1972, best characterised by pollution control. After all, one of the *Limits'* five parameters limiting growth was pollution which at the time was seen as an inevitable companion of industrial development.

Let us therefore have a closer look at the success story of pollution control which was the first success in the "decoupling" of parameters that had been seen as mechanically coupled.

Industrial pollution was one of the most conspicuous negative side effects of economic growth during the 1960s and 1970s. Its impact was largely localized. A suggestive kind of "solution" was its "de-concentration" using higher smoke-stacks or other distribution technologies. That, however, didn't solve the under-lying problems, as *Limits to Growth* clearly demonstrated.

Later, however, emission control technologies set in and were made manda-tory in the industrialized countries. Standards were set by law and licensing pro-cedures adopted for industrial installations. Eventually some international har-monization was achieved for pollution control standards, notably within the European Union. Pricing instruments were also introduced throughout the Euro-pean Community and in other OECD countries to make the polluters pay.

Pollution control became a marvellous success story. Rivers and the air in in-dustrial agglomerations became clean again, in the wealthy countries, that is. The graphic representation for this success story was developed: the "inverted U-curve" or environmental Kuznets curve with pollution plotted against time and economic development. Countries typically start at the lower left corner, poor

---

1 Meadows, Dennis, Donella Meadows, Jorgen Randers and William Behrens. 1972. The Limits to Growth. New York: Universe Books

and clean, and then they succeed in industrialisation and become rich and dirty. Finally, when those nations are rich enough to afford pollution control they end up rich and clean.

This experience allows you to say it is good to be rich so that one can afford costly pollution control. Or, with a slightly modified meaning you can quote Indira Gandhi's famous statement made at the first UN Conference on the Human Environment, in Stockholm 1972: "Poverty is the biggest polluter". This famous statement went down extremely well with developing countries because it allows them to go on with traditional growth strategies and claim that the pollution is finally good for the environment. The OECD countries are in love with the environmental Kuznets curve, for the same reason.

Pollution control is also a favourite in the business community for example, in business communication and in Corporate Social Responsibility. But it is not only industrialists and developers who adhere to the idea that environmental policy is best represented by pollution control. Environmental professionals both in the public and private sectors tend to like it because most of them are trained in pollution control anyway, whether as technicians as administrators or as lawyers. So much for classical environmental policy.

Coming back to the challenges to sustainable development, it has to be said clearly that the inverted U-curve does not automatically apply to all environmental problems. To begin with, it wouldn't have applied to pollution in the absence of strict legislation. The main question is whether other relevant environmental factors such as greenhouse gas emissions, habitat destruction and unsustainable life styles can also be de-coupled from human well-being. At this time they are not de-coupled at all. Almost the opposite of Indira Gandhi's statement is true for these factors. Hence, prosperity assumes the role of the biggest polluter.

This is so embarrassing a phenomenon that economists and politicians prefer not to recognise the realty. They are quick to involve *sustainable development* which says that economic and social well-being are also indispensable and who dares to contradict? Often, notably in the USA, politicians and business leaders swiftly return to the comfortable and familiar paradigm of pollution control.

A typical example for this psychological repression of reality is the new hype for fuel cells. Their great advantage is clean air at the local level. But if the *primary* source of energy remains unchanged, as can be expected in our days, fuel cells have no advantage of present technologies as regards climatic change.

How can we politically address such topics as climate change, biodiversity losses and unsustainable lifestyles? Chiefly, I suggest by truly *de-coupling* wealth from what has been identified as the ecological problem.

In the case of the greenhouse effect, what we need and what we can achieve is a dramatic increase of energy productivity, meaning that we extract twice or four times or ten times as much wealth from a barrel of oil or from a kilowatt-hour as we do today. Renewable sources of energy also come into the picture. For biodiversity protection, we have to learn to use less space for our well-being so that more space remains for wild plants and animals. Sustainable lifestyle means prosperity with low consumption of non-renewable resources.

With regards to policy instruments for our new task of de-coupling, we can learn from classical environmental policy. Let us look at particular at pricing instruments.

For pollution control, pricing instruments have greatly helped cleaning up the environment. A typical case has been the waste water charges whose revenues were used to finance water purification. This system of charges was never very controversial. It fully conformed to the polluter-pays-principle and it left those who successfully applied prevention measures essentially free from the charge.

In a broader sense, the same applies to user fees, refund systems, violation penalties and tradable emission permits for classical pollutants such as SO2 or NOx. They too met with little public resistance when introduced.

In the case of long term and global environmental problems, notably the greenhouse effect and life style changes, pricing instruments will have to work on input rather than on output factors. Here, completely avoiding the fees or taxes is hardly conceivable. Nevertheless, if efficiency gains more or less corresponding to the price increases for energy or other scarce resources, the pricing instrument need not be very unpopular.

I am confident indeed that a new universe of eco-efficient technologies will become available once prices for the use of scarce natural resources begin to reflect their long-term scarcity. At the Wuppertal Institute for Climate, Environment and Energy we have sketched out the roadmap towards that new universe. Together with Amory and Hunter Lovins I published a book, also as a Report to the Club of Rome, and called it "Factor Four – Doubling Wealth, Halving Resource Use"[2].

The book presents fifty examples, from automobiles to household appliances, from buildings to logistics, from industrial processes to farming methods, all demonstrating that a factor of four is available in energy or material efficiency.

The factor four model can be seen as most promising to everybody dealing with climate change, urban sprawl and biodiversity losses.

---

2 Earthscan, London, 1997, also available in other major languages including Chinese and Japanese.

But there exists another difference between the factor four concept and the classical pollution technologies. The development of pollution control technologies took something like ten or maximum twenty years. The full implementation of the factor of four model for achieving the necessary resource productivity will take fifty years. And a socially reasonable and comfortable reduction of urban sprawl may take a hundred or two hundred years.

The long time frame can also be seen when looking at price elasticity. Fuel consumption of the car fleet will not change swiftly upon a small price signal. The immediate price elasticity is low. However, if society knows that energy and other resource prices will go up slowly but steadily for a long time with no hope of their coming down again, auto manufacturers will strategically invest in fuel efficient cars. Customers will develop a preference for fuel efficiency. Engineers and scientists will target the basics of resource productivity. And public planning will shift priorities towards convenient mass transport and agreeable high-density urban planning. Similarly, other environmental challenges will be faced and answered over time. High energy efficiency in buildings, lower transport intensity for manufacturing and food, a growing share of renewable energy and a systematic reuse of materials in the production cycle and after use will become technological mainstream. A factor four model becomes a realistic perspective for all sectors. In other words, we can expect high price elasticity in the long run.

Long term price elasticity means that price signals should be mild but predictable. A political all-party agreement over thirty or fifty years to raise prices for scarce resources in very small and predictable steps, preferably in steps so small that technological progress can keep pace, would be best.

Please note that I am talking about a *price* corridor, not a *taxation* corridor. Taxes or other instruments would be used to stay inside the price corridor. In this ideal case, the monthly bills for petrol, electric power, water, space or virgin raw materials remain stable. On average, the population would not suffer any losses in their lifestyles.

If the fiscal revenues from this operation are invested in reducing indirect labour cost, you would expect positive effects for employment. Compared to 'business as usual' scenarios, you would see human labour services becoming gradually cheaper, for example, more affordable for the clients.

Ultimately, the factor four technologies will be seen as revolutionary as those characterising the Industrial Revolution from the nineteenth century onwards. The difference is that the Industrial Revolution was all about the increase of labour productivity while the new paradigm is about the increase of resource productivity. There are no physical or technological reasons prohibiting a similar increase of resource productivity, but it can also take two hundred years.

World History from the Viewpoint of Sustainable Development

Karin Feiler and Gabriele Zöbl

In order to get a feeling for future prospects, we must look back deep into the past, says Aurelio Peccei in his book "One Hundred Pages for the Future," published in 1981. For it is only from a broad perspective that we will see our generations assume their rightful place at the pinnacle (at least to date) of our evolution. According to Aurelio Peccei, we must go back very far indeed to have a complete overview of our development – even though conditions throughout the course of world history have always varied greatly, and the solutions applied often led to development that was not sustainable and that was brought back into a balance beneficial to humans and their surroundings only by applying countermeasures.

Beginnings of Mankind – Equality with No Concept of Property

Social and Environmental Aspects:

Human development has always been connected with the development of the climate. The discovery of the skeleton of a prehistoric man, also named Millennium man, in the year 2000 in Africa let us suppose that most likely between five and seven million years BC Africa was already a bountiful land. But the development of mankind took million years until, *Homo erectus Modjokertensis* (Java man) appeared in Java and *Homo erectus leakey* lived in Africa. The first human beings roamed the tropical jungles in small tribes, digging roots and hunting deer to meet their food needs. The next phase of evolution saw the arrival of *Homo erectus erectus* in Java and *Homo erectus mauritanicus* in North Africa. The climate allowed the inhabitants to live without any concept of sheltered dwellings. The first inhabitant of Germany, Heidelberg man, appeared some two billion years ago and had a similar lifestyle. In addition, archaeological findings indicate that around the same time, Peking man in China was the most highly-developed of all humans. Even during the interglacial period between the Mindel and Riss Glacial Stages, approximately 350,000 to 200,000 years ago, Steinheim man lived in Europe that was still characterised by a mild climate.

In contrast, his successor, Neanderthal man, had to contend with an Ice Age. Thus from approximately 150 to 80 thousand years ago, the Neanderthal's original homeland was a deserted, icy tundra. This varied only in Southern Europe, which consisted of grass steppes and scanty taiga forests. Neanderthals had to disperse, probably due to the cold and lack of fertile ground, to areas in Siberia, Africa and the Middle and Far East. Their life expectancy was well below 50 years, with about 40% dying before the age of 20 and an additional 40% dying between the ages of 20 and 30. Neanderthal man became extinct sometime between 40,000 and 35,000 BC.

In the Americas, the first settlement started in about 40,000 BC. In all probability, this occurred via migration from eastern Siberia across the Bering Sea, which at that time was still a land bridge to Alaska. This human dispersal continued southwards; Venezuela was not settled until about 14,000 to 12,000 BC.

At the same time, the Cro-Magnon people appeared in Europe. They were excellent hunters during the decline of the Ice Age, and also had many of the same characteristics as today's *Homo sapiens sapiens*, such as coherent language and the ability to make tools. There is archaeological evidence that the Cro-Magnon people built tents made of wild-animal skins and also settled in caves or in overhangs. The evidence of drawings on cave walls also highlights the beginning of artistic thinking. People lived in tribes consisting of 15 to 30 members.

Economic Aspect:

The art of tool making was passed on from one generation to another. The small human and animal populations in an environment with a good climate resulted in self-sufficiency and even an abundance of goods from nature, including food supplies. In light of this dependence on nature, group solidarity and economic relationships probably consisted of barter and exchange rather than joint production.

In summary, from today's perspective, the beginning of human development can in fact be called sustainable, since humans took from nature only as many resources as were needed to ensure their survival. Finding food was very hard work; life expectancy was further significantly reduced by wild animal attacks. From Nature's point of view, this epoch was certainly to be judged a success. Humans were subject to Nature.

They did not have the strength or power, either from a technical standpoint or as derived from their small social units, to intervene in the "course of nature".

The New-Stone-Age (Neolithic) Revolution – Beginning of Agriculture and Social Structures:

Social Aspect:

The New Stone Age (Neolithic period) lasted from 10,000 to about 3000 BC. Before, during the Palaeolithic Age, the old hunter-gatherer units developed into farming villages which became the new settlement units. At the beginning of the Neolithic period, typical villages already counted some hundreds of inhabitants and towards its end, some cities already had a population of several thousand. For instance, the ancient Sumerian city of Ur had a population of 34,000 people on 89 hectares of land around 2800 BC. These villages developed a differentiated social system, which evolved from an originally egalitarian society into one with many different social strata.

The core question is whether the emergence of agriculture resulted from human desire to form clans with fixed power structures, or was it the survival need of a rising population to establish clans to secure the food supply from the nearby fields, instead of hunting animals and gathering plants from the surrounding areas.

The British Scientist, Gordon Childe, has postulated that agriculture emerged due to dramatic climate changes following the Ice Age, which resulted in the concentration of the population, especially in the most habitable areas, i.e., those that were not drained and desolate. Thus, an increase in food scarcity was avoided.

Environmental Aspect:

This Neolithic revolution was centred in the Middle East, Mesopotamia, Egypt and the eastern Mediterranean, all beyond the boundaries of the ice areas to the north. Agriculture was in a sense a precondition for the development of complex social structures and technological innovation, including the production of tools.

Economic Aspect:

Around 8000 BC, the first systematic crop cultivation took place independently in the Middle East, East Asia, Mexico and Peru. The crop varieties grown clearly differed from the wild types. In contrast, collection and domestication of these varieties by the Neolithic people did not happen through direct intervention, but rather through a process of adjustment and unconscious selection.

Through constant repetition of this unconscious selection, millet was the first plant to be domesticated, around 4000 BC. In Thailand, field beans and a type of pea were domesticated about 7000 BC, and rice was also being cultivated by around 3000 BC, well before it was grown in China. Between 5200 and 3400 BC, the Mayan and Aztec cultures used maize as a staple food, long before pumpkins and beans were domesticated. In the high land of Peru, which was inhabited by hunter-gatherers as early as 15,000 BC, cultivation started around 5600 BC, but systematic crop cultivation did not begin until 2500 BC.

Soon an adequate food supply was available in all parts of the world, including wheat, barley, oats or lentils in Europe, and pumpkin and avocado in America. In East Asia, almonds, cucumbers, or millet dominated; in China, the latter was even more important than rice until the second millennium BC.

At the same time, animals were domesticated to meet the people's needs. From 6500 BC on, domesticated goats and sheep could be found north of the Persian Gulf. According to Cyril Dean Darlington's book, written in 1971, mixed agriculture began between 8000 and 6000 BC. Between 6000 and 3000 BC, agriculture was divided into the farmers and the stockbreeders.

The surplus of agricultural production permitted specialisation and a division of labour in different activities, such as building irrigation systems or defences, or organising independent groups of priests and warriors.

To summarise, it turns out that the Neolithic period was a very crucial phase for humanity. Settling down in village-like communities and beginning to farm significantly increased physical safety. At the same time, a social hierarchy was created within individual kinship groups. Intervention in nature occurred through slow adaptation and unconscious selection, not through direct interventions as with conventional grafting or genetic engineering.

Advanced Civilisations: Economy and Government under One Roof

Social and Environmental Aspects:

At the end of the Neolithic, around 3000 BC, advanced civilisations began drastically to modify the social hierarchy. The structure of kingdoms was developed. In Egypt, one of the first advanced civilisations (3000 to 332 BC), the king was regarded as a divine being. In Mesopotamia the whole country belonged to the town god and all administration, economy and culture were run by the priesthood; the king was appointed as the divinity's terrestrial representative.

After thousands of years of evolution, from the Stone Age to an era with a coherent language and culture, human development had reached its peak in the advanced civilisations. These cultures already had written law (Roman law actually still serves as a partial basis for European civil law). A whole range of professions was started, such as those of scribes or various craftsmen. The first craftsmen created more detailed works of art and scientists reached a high level of mathematical and technical knowledge. However, these achievements should not divert attention from the fact that in Egypt, Mesopotamia and Rome, humans had to perform hard labour as slaves every day. They had no legal rights. Under Roman law, slaves were seen as objects (*res*) and had the same standing as animals.

Thanks to its superior iron weapons, Assyria became the new dominant power in the Middle East. It was not until 322 BC that Alexander the Great was greeted with cheers when he liberated Egypt from the foreign rule of the Persian Empire, which had risen to power in the meantime. In 395 AD came the split between the Western (Roman) and Eastern (Byzantine) Empires.

In all empires, wars weakened the economy, which led to domestic tensions and strikes. The earliest known strike was called in 1156 BC, in Diirel–Medine, Egypt, because in-kind payments to the work force were not made. Even though reforms were instituted, they were not enough to prevent the decline of Egypt's power under Ramses III in 1153 BC. In Athens, in 594 BC, Solon ordered reforms for the city. Among other reforms, he abolished slavery and cancelled farmers' mortgages.

In Rome, around 450 BC, the Law of the Twelve Tables strengthened plebeians' rights. However, the unreasonable conditions for slaves later (in 132 BC) led to the four-year Sicilian slave war in the Roman Republic, in which 70,000 slaves were involved.

In 62 AD, Pliny the Elder was already reporting first serious environmental degradation from air pollution in Rome, caused by fireplaces.

Economic Aspect:

With the rise of advanced civilisations and the emergence of trades, the concept of labour as service in return for compensation developed. Of course, it took another few millennia, until the Industrial Revolution, prior to which workers generally received only room and board for their services.

The consideration of war, in order to defend the property, led for the first time to a new craft: weapons manufacture. National economies existed on the lowest level, based on the barter of agricultural goods. The first coins were introduced in 1200 BC in China and 700 BC in Asia Minor.

Over considerable periods of time and in various respects, including from the standpoint of the contemporary world view (as shown by the many domestic tensions), economic, social and environmental developments were not to be regarded as socially acceptable and therefore not as sustainable. Economic consolidation was always significantly promoted by the introduction of trade and the expansion of the trade area. The poor social conditions under which the underpaid, dispossessed work force or slaves lived cried out for structure-altering political reforms such as, for example, those implemented by Tiberius Gracchus' land laws in the Roman Republic or Solon's constitutional reform in Athens. These social measures represented a further step toward sustainable development. In addition, due to the clearing of forests for shipbuilding, environmentally-related countermeasures would have been necessary.

Europe in the Middle Ages

Social and Environmental Aspects:

The European political situation changed considerably in the Middle Ages, after the fall of the Roman Empire in the fifth century. Mass migration had already begun in the fourth century. The Visigoth, Ostrogoth, Frankish, Burgundian and Hunnish tribes left their homelands. The Teutons had already been on the move for a long time. They brought the Roman Empire to an end when they dethroned the last emperor, Romulus Augustulus, in 476 and replaced him with Odoacer, a Teutonic army commander. In 511, the Merovingian king Clovis I established the Frankish empire. By the eighth century, the Carolingian dynasty had already succeeded the Merovingians, the high point being Charlemagne's coronation as Emperor in 800. The Holy Roman Empire consisted of the kingdoms of Germany, Burgundy and Italy. Later, beginning in the eleventh century, it no longer had a central authority. Political emphasis shifted to individual aristocratic dynasties. Elected monarchs were the rule of the day. In contrast to the Holy Roman Empire's rulers, the West Frankish kings were successful as hereditary monarchs until 1200. Transfer of power was characterised by the constant turmoil of war, as with the replacement of the Guelph kings by the Hohenstaufen King Frederick II, the accession of the Capetians in France in the thirteenth century, or the Hundred Years War (1339–1453), when the French, under Joan of Arc, defeated the English at Orleans and the French nation was born. In 1453, the Muslim Ottomans conquered Constantinople, the capital city of the Byzantine Empire, ending the more than one thousand year history of this successor to the Roman Empire. The Middle Ages are considered to have ended in 1492 with the discovery of America. At that same time, the idea that the Earth is round and the sun is at the centre of the universe was slowly winning out. In 1492, the year America was discovered, Martin Behaim of Nuremberg designed the first globe.

Life in the Middle Ages was characterised by the rise of Christianity. Other influences on the Germanic world included the Latin language, portions of Roman law and Germanic ideas and ways of life. The feudal society characteristic of the Middle Ages originated in the way noblemen were paid for performing military service. The king would present them with fiefs (Latin *feudum*), which were worked by the peasants "bound" to them (bondservants). In compensation, the peasants and serfs received protection and the right to room and board. The social hierarchy subdivided into the nobility, freemen and bondservants.

The Middle Ages were marked by the religious control exerted by the Roman Catholic Church. The Church reached the aggressive pinnacle of its power in the thirteenth century with the Inquisition, and its politically weakest point with the schism in 1378. Church and State were inseparable. Any attempt at religious, social or political reform was necessarily directed against both and was hunted out by both together. In 590, Pope Gregory I succeeded in becoming the secular ruler of Rome, thus laying the foundation for the creation of the Papal States. Monasteries founded in the seventh and eighth centuries were setting up an ever-increasing number of schools based on the Eastern model. Since enrolment was not restricted solely to the nobility, schooling represented the only possibility for social advancement for many people. The first university was established in Bologna in the mid-twelfth century.

Social life was characterised by the creation of cities, sometimes in the former locations of Roman cities, which had fallen into nearly complete disrepair, and sometimes as diocesan towns. In the rural world, the rights of freemen and bondsmen became more similar. From the serfs' compulsory associations, cooperatives arose in the twelfth and thirteenth centuries that had greater autonomy where, for example, use of a community's agricultural land according to the already well-known three-field system was concerned. The increase in the number of money transactions made it possible to pay off heavy obligations. However, most peasants remained in bondage and owed service and taxes to their feudal lord.

About 1224, the *Sachsenspiegel*, the oldest written law book from the German Middle Ages, soon achieved a level of importance extending far beyond the empire. Until then, all of the Germanic tribes had adhered to orally transmitted common law.

The agrarian economy was still keeping pace with population growth. In central and western Europe alone, the population grew from about 5.5 million ca. 650 to about 35.5 million in 1340. The crisis came in the late Middle Ages, beginning in the middle of the fourteenth century. The great plague descended on an empire that in the meantime had become overpopulated and was racked by famine. By the middle of the fifteenth century, over a third of the population had died. As the fifteenth century waned, this period ended with misery and great religious movements that weakened the Roman Catholic Church. However, cultural life attained special significance with the invention of printing in 1447. Humanism's fine arts and those of the Renaissance should not go unmentioned. The great voyages of discovery are considered to be a defining event of the period, and to mark the end of the Middle Ages. The year 1492, in which America was discovered, can be seen as an especially noteworthy moment in time.

Economic Aspect:

As new cultural areas became known, first through the Crusades and later through continually increasing trade, people gained a broader conception of the world. By the thirteenth century, the merchant confederation of Hanseatic cities such as Hamburg and Bremen had already been transformed into a confederation of cities that could represent their common interests and protect trade privileges in foreign lands. The basis for a global economic area was created, at the latest, when Australia's north-west coast was discovered. In the absence of insurance, attempts were made to reduce the risks inherent to sailing vessels somewhat by distributing them among ship owners through risk pools.

From a national economy standpoint, tax policy was conducted by each individual ruler by collecting property taxes in kind or in money. At the level of businesses, mediaeval economic policy measures appeared in the flourishing cities, particularly with the founding of guilds, which had their roots in the Roman Empire or could be traced back to brotherhoods and associations of a religious sort. Organisations of traders or craftsmen controlled, among other things, product prices and quality, the number of apprentices and the limits on the numbers of craftsmen. Social policy was left up to the individual or to local authorities.

Seen as a whole, the European population of the Middle Ages had access to a high level of potential resources. Regular seasonal changes, extensive forests and important raw materials, such as ore, formed a good basis for continuous economic growth until the thirteenth century. However, an individual's ability to make a living was, on the whole, dependent on the class into which he was born, and his life was greatly affected by the many turmoils caused by war. Living conditions undoubtedly varied greatly. Due to rapid population growth, the weakening of the Church's spiritual and political supremacy, wars, and the onset of the plague, the late Middle Ages in Europe, beginning in the fourteenth century, appear bleak. The turning point came thanks to an intellectual revolution, the flowering of the arts and the opening of Europe that accompanied the great voyages of discovery at the end of the fifteenth century. Our yearning for new things allows us to conclude that, since the beginning, humanity has always had to seek out new and undiscovered things, in order to be able to safeguard its physical existence through the continual intellectual development of new horizons. So it was that afterwards, the changes always led to crucial changes in the basic political conditions. Drastic interventions in Nature were no longer something to be stopped; however, at that time they had only local effects.

A New Era – The Discoveries – Globalization in its Infancy

Social and Environmental Aspects:

The oldest written report about a discovery is the one about an expedition to Punt commanded by the Egyptian queen Hatshepsut at the beginning of the 15th century BC. All important discoveries before the birth of Christ concerned either Africa or the Middle East. The conqueror Alexander the Great even went as far as the threshold of India. The largest expansion of the Roman Empire was achieved under Trajan (98-117 AD) and encompassed a huge area stretching from England to Mesopotamia.

The seafaring Vikings, the first to turn westward, reached the north-east coast of North America around 1000 AD. However, most of their knowledge of America was lost. For a long time, the Arabs, using geographical maps from the Greeks, possessed a far better knowledge of Africa, the Middle East and South Asia than the Europeans. In the 13th century AD, knowledge of Asia took a quantum leap thanks to the papal couriers and Marco Polo, who discovered the Pacific Ocean. The new era started with the discovery of America by Christopher Columbus on 12 October, 1492, during the search for a westbound route to India – a search initiated in part by the fact that in 1453, the Turks had cut off access to India at Constantinople.

The voyages of discovery that were just beginning as the fifteenth century drew to a close were followed by heavily-financed expeditions in the sixteenth century. Japan was discovered, Magellan became the first to successfully circumnavigate the globe, and Francis Drake (1577-80) the second. As the leading nations of the time, Spain and Portugal lay claim to territory discovered by them.

With the start of the 17th century, the English (1600) and Dutch (1602) created commercial companies and established the basis for their colonial empires in India and the Malaysian archipelago. The final cornerstone of a global market was laid with the discovery of the NW coast of Australia on 28 January, 1788.

Through determined policy and war, but also through marriage and the politics of inheritance, the Habsburgs continued to extend their dynastic power from the fifteenth century on. The foundations for the Habsburg Empire were laid under Maximilian I (1439-1519), and the dynasty remained in power until the end of the Austro-Hungarian monarchy in 1918.

A highlight of religious history came in 1517, when Martin Luther nailed his 95 theses on the Wittenberg castle church and ushered in the Reformation. In 1532, Charles V enacted the first imperial criminal code, but it was not until the mid-eighteenth century that witch-burning was gradually abolished.

In the 16th century, the first chimney hood and condensing chambers were developed in Bohemia and Saxony to traplarge particulates arising from the iron and steel industry.

Economic Aspect:

In 1798, British pastor, national economist and social philosopher Thomas Robert Malthus anonymously published his work on the causes of population increase. In the face of limited food reserves, Malthus propagandised for birth control. For Europe, the ideal long-term solution to its problems came from America. Although potatoes had already been cultivated in 1786 under Frederick II, the Great, it was over 150 years before they were used as a new staple food. The yield of potatoes per hectare was nearly three times as high as the yield of cereal crops. However, many people died because they consumed the poisonous fruits instead of the potatoes. Others cooked only the leaves.

Humanity got off to a complicated start in modern times. On the one hand, there were new discoveries, on the other, large-scale religious wars. During these centuries, the focus seems to have been on expansion and on combating Europe's recurring food shortages. Scientific and technical progress came about rather slowly. The invention of celestial mechanics by Johannes Kepler in the seventeenth century is especially noteworthy.

Industrial Era

Social and Environmental Aspects:

In the eighteenth and early nineteenth centuries, wars within individual European countries were prevalent on the Continent. Austria hosted the Congress of Vienna, intended to re-establish order in Europe after the Napoleonic wars. Slowly, constitutions were introduced in the various countries. The educational system flourished once again as many new universities and colleges were established.

Social conditions became noticeably worse as industrialisation proceeded. With the emergence of textile mills, peasants – those in Great Britain first – lost their only income source, namely spinning wool. Agricultural workers were forced to leave their farms and move close to the cities, i.e., the mills. The constantly increasing numbers of people moving into the cities led to impoverishment of the population. The associated aggravation of social tensions led to protests in Great Britain as early as the end of the eighteenth century. In both Germany and Great Britain, the late 1840s saw the creation of professional trade unions, such as those for printers and cigar workers. At first, these associations were concentrated in the manual trades, in which guild traditions usually continued. In the US and France, unions developed in the late eighteenth century. In Austria, the political repression of the Metternich era kept any efforts to form unions in check until the March revolution of 1848. Even after that, starting a union was prohibited until the end of the nineteenth century. In Germany, the free (i.e., socialist) trade unions developed into a mass movement that lasted until 1914. In 1933, the union houses were dissolved and forcibly incorporated into the *Deutsche Arbeitsfront* (German Labour Front).

The industrial revolution also brought changes to the social support system. As early as the first half of the nineteenth century, once again starting in England, all of Europe began to enact legislation to protect workers. Such laws set about reducing work time for women and children, improving sanitary conditions in the workplace, and instituting protection against accidents.

However, obligatory coverage of factory workers by compulsory insurance was not attained until the German Reichstag (parliament), under Bismarck, enacted laws in 1883-1889 providing health, accident and disability insurance for workers. No European country remained unaffected by what Germany had started. Austria followed suit in 1887 with workplace accident legislation and in 1888 with a law covering worker health insurance. It was not until 1911 that England decided to provide medical treatments for blue-collar workers and low-paid white-collar workers, with the Netherlands doing the same in 1913. France, Italy and Sweden were unable to stomach the thought of social health insurance until the period between the world wars. England, Denmark, Sweden, Italy and Belgium were the first to introduce protection for the elderly and the disabled. At the dawn of the twentieth century, in 1911, England became the only European country to cover the risk of unemployment through statutory insurance protection. On the American continent, unemployment insurance did not catch on until after World War I.

During the First World War, 200,000 aircraft were already being produced, although brothers Orville and Wilbur Wright had only just acquired a good grasp of the technology needed to control a plane, in 1900. In the 1930s, increasing amounts of money became available for research and development at the same time that R & D was developing into a separate professional branch.

Air pollution of human origin resulting from combustion process, further increased with the invention of steam-driven machines.

Economic Aspect:

What happened? In 1764, the invention of the world's first spinning machine by hand-weaver James Hargreaves from Stanhill brought manual labour's dominance to an end. However, this spinning machine, known as the spinning jenny, was suitable only for producing softly turned yarns. A set of further technical innovations brought the breakthrough for the British textile industry which, around 1750, employed 27% of all workers in the wool business. The spinning of wool remained a hand operation until 1779. A few years later, home weaver Samuel Crompton invented the spinning mule from the water frame spinning machine designed by Richard Arkwright (1768). The machine was first manually controlled, but by 1790 it could be propelled by engines. The Briton James Watt designed the first double-action steam engine (patented 1760) based on the low-pressure steam engine invented by Thomas Newcomen in 1712. It was utilised industry-wide by iron workers in the coal mines. The idea of moving a piston with steam pressure had first been raised in 1690 by the French engineer Denis Papin, the inventor of the pressure cooker. In Great Britain, beginning in 1760, iron ores were smelted in coke-fired blast furnaces developed by Abraham Darby, who had utilised these concepts since 1709 in Coalbrookdale.

Adam Smith saw in the Industrial Revolution the key to ending the eternal fight of mankind against the scarcity of goods and to reaching the target of prosperity, through the transition from an agricultural economy to an industrial one. In 1776, he laid the foundation for economics as an independent discipline. This science was to be seen as separate from the numerous past observations of prior economies. This opinion was not undisputed even during Adam Smith's days.

Quesnay, the French economics economist, who was famous for the *laissez-faire* policy, considered agriculture to be the most important source of wealth.

The further development of production processes focused on progressive industrialisation in order to achieve higher productivity: "more goods with fewer resources (less time, less labour and decreased material inputs).

It was not only stronger trade with the new manufactured goods that increased the demand for money. Industrial technology had moved to centre stage in the struggle to increase wealth and welfare. Capital investment was necessary to produce the technology required for the new quantitative leap forward, in order to achieve the required increase in efficiency. Movement of the labour force toward the new production facilities was a given: the modern manufacturing plant was born. This meant higher efficiency due to installation of machines in factories and the mobility required of the workers. The focus was on the specialised production technology and ever more efficient, thus faster and less manual-labour-intensive machines.

Industrial technology had thus become the focal point in the battle to increase wealth and well-being. The technological leap at the beginning of the Industrial Revolution was not a qualitative, but a quantitative one. Technology had been around ever since man started doing handicrafts, but the real technical challenge came later. The problems of inventing materials and related mechanisms had to be solved.

All in all, the change in world events during the Industrial Revolution can be viewed as the beginning of our current world view, in that it divided the world into industrialised and non-industrialised countries. Sectoral policies such as economic and social policy came into being. From the beginning of the nineteenth century on, technological progress proceeded at breathtaking speed compared to previous centuries, and was accompanied by socio-political action. The first signs that humanity was on the verge of seeing economic growth as being constituted solely by a continual increase in the quantity produced were crystallising even as the very first factories were being established.

The "Golden Quarter-Century"

Social Aspect:

After the end of the Second World War, Europe owed its rapid reconstruction to the unflagging diligence and application of its citizens. On the one hand, considerable payments had to be refunded to the winning powers; on the other, the Marshall plan supported economic development starting in 1948. Most of the work was done by women, who had already been involved during the war as back-up workers for the male work force in the factories. Many men were still prisoners of war, and women dominated the social picture. They also had to collect the rubble of their houses so that the materials could be used for rebuilding.

Environmental Aspect:

Environmental pollution was quite easy to identify. Its sources were plants from the heavy industry, which were already operating during the Second World War. The potential of the newly built highways in Central Europe was still almost untapped, but available for the new mobility of the future society. The first cars for the general public were already being produced.

Economic Aspect:

Only towards the end of the 19th century were new tools and products manufactured based on scientific knowledge, i.e., on the examination and understanding of problems beyond the immediate perception of our senses.

Until the mid-1920s, investment in research laboratories was not at all consistent. Of course the dramatic economic breakdown after the First World War needs to be recognised.

The cost of production could be accounted for only in terms of the cost of labour and capital. Decisive investment in research and development started in the 1930s. From then on, these activities were seen as professions in their own right. Today, a company has to invest 25 to 30% of its total sales income in the research and development of a new product before the product actually goes to market. This takes 10 to 15 years. A big change in modern times finally coupled technological applications with the advance of scientific knowledge after the Second World War.

After the second Word War the so-called industrialised states in Europe experienced a miracle of economic growth. A continuous increase in growth rates persisted for 25 years. In terms of quantitative economic growth, this has been a unique phenomenon in the history of mankind. It ended in the 1970s.

What was newly invented? Highways, aircraft, traditional industry, radio and television.

Steady industrialisation has been recognised further on as key to future development, in order to combat a shortage of goods efficiently. Resources have always been treated as an inexhaustible fixed element of this earth.

44

But after the Second World War, for the first time in the history of mankind, the value added portion shifted from the production sector to services. The age of the service society was born.

Economic considerations centred on the condition that production has to be efficient, always in line with the quantity of units produced over a certain period. The cost/benefit ratio was calculated only by directly comparing production costs and sales prices.

Neo-classic economic theory enlarged the narrow classic labour value theory characterised by offer and production, and led to speculations as to whether the real overall product should be increased in relation to the demand. This process also revealed that demand is based on benefit. But economics did not react to the demands of society in an ever-changing world.

When the Club of Rome published its report entitled "Limits to Growth" in 1972, shortly before the oil crisis began, the idea of limited economic growth struck western industrialised countries like lightning out of a clear blue sky.

Thanks to technical advancements in the fields of energy, efficient use of resources, and hydrogen technology in recent decades, the twenty-first century economy finally makes it possible, from a technical standpoint, for individuals to choose between a sustainable or non-sustainable way of life. Whether or not they can afford it depends on basic political conditions. This aspect depends on policy. This is a signal that no one can evade responsibility.

Through our overview of events, we have arrived at the knowledge that epoch-making structural changes in the economy, the environment, and social life required always an amount of time that seemed long for the respective period. With progress, however, the intervals became shorter. In consequence, the world is now changing at an enormously accelerated rate. While it took about three thousand years to develop agriculture, it took only a decade for the Internet to become widespread.

The following chapters will take up the situation in today's world and suggest solutions, in the various areas of life, for ways in which sustainable world-wide prosperity can be achieved in the future, within the context of the sustainability model.

III. The Most Influential Factors on the Global Development of Environment

The Demographic Development

Josef Schmid

⇒ World population will increase to around 9.3 billion poeple by the year 2050
⇒ The basic requirement for a standstill of the population growth, families should have two surviving children (an average of 2,2)

World population is double-faced. On the one hand, it appears as the sum total of humanity that currently populates the globe with 6.2 billion people. On the other hand, world population is the combined inner life of the 200 states and public entities of various sizes distributed all over the globe. The world population aspect as such, i.e., regardless of state borders, is generated by a supranational interest, usually a concern about the inhabited planet and its future. The strong feeling of unease caused by the world population issue ever since the last century has its roots in two causes: its hitherto unprecedented growth and the lack of any defined responsibility for the world population as such, as only individual states bear concrete political responsibility for "their" populations. If the world population, in particular their growth, were to be given a new direction, such as by curbing growth, this would be the sole business of the states with the largest and fastest growing populations that would have to be influenced depending on their state of awareness.

The United Nations, founded at the end of the Second World War, have striven to lessen this discrepancy in which world population finds itself, i.e., being a global problem while at the same time coming within the subdivided responsibility of sovereign states. In 1946, they created a special research and observation division, the *Population Division*, which annually submits to the public a *"Demographic Yearbook"* and regular future projections on development and growth of the world population.[1] The "World Population Conferences", organised every ten years by the United Nations Economic and Social Council, with their adoption and regular review of "World Population Action Plans", have finally led to the emergence of a global public that creates awareness of the world population growth as a whole and of the responsibility borne by the individual

---

1 United Nations, World Population Prospects – The 1996, 1998, 2000 Revision, New York; United Nations, Demographic Yearbook, New York (published annually), United Nations, International Migration Report 2002, New York.

states with regard to "their" population problem that has an altogether trans-boundary impact.

After the end of the Second World War and the rising tension between East and West, the population growth of the "colonial peoples" – all striving for independence, which they were soon to achieve – , became a matter of growing concern. The emphasis was on the ideological and geopolitical question which political system they were going to join. Seeing large populations like the Chinese drifting into the Communist camp, India's and Egypt's seesaw policy between blocs, were highly alarming. In fact, the world population problem has always been localised beyond the modern world and seen from a strategic angle by the hegemonic powers, the USA and the Soviet Union.

The 1972 Report "Limits to Growth" to the Club of Rome forced a change of paradigms, despite the deficiencies and shortsightedness of this pioneering work. The growth of large populations, unknown in European history, is detached from the pure power context of global ideological rivals and put into relation to the basics of human subsistence. The Report placed "growth" of both economy and world population on the test stand of a world whose resources are finite and limited. Ever since, the demand for a balance of the number of people and their material needs has been on the table, supplying the keynote of pertinent scientific works ("zero growth") and international conferences on environmental and population issues. The end of the East-West conflict, which also put an end to an ideological century, finally opened up an unobstructed view on the "One World", the "Blue Planet", i.e., the biosphere as the only place of human existence.

It is due to this development that the notion of a balance with regard to humanity and environment has achieved an international obligation with the sustainability concept, which was hardly to be expected in a world of material rivalries. More than two decades ago, Lester Brown's "World Watch Institute" made "sustainability" the pivotal point of global satisfaction; the breakthrough came with the world environment report ("Brundtland Report") which defined sustainability as the next generation's chance for development that must not be obstructed or diminished.[2]

Population within a conception of sustainability can only be perceived as a system conception where the modification of one element generates equilibrating or correcting responses of all other system elements. Population changes that cause responses in a system of sustainability are growth or shrinkage with the entailing shift in the age structure. Active age groups between 15 and 65 need an occupation, dependent young age groups – education and training, while old age

---

2 World Commission on Environment and Development. Our Common Future. Oxford University Press, Oxford, UK, 1987.

groups need old-age protection. Demographic changes multiply by the demands of individuals to their existence. They are linked to cultural tradition and the stage of development. A rising consumption of financial or material resources must be compensated with an enhanced quality of the technology used and new forms of organisation to ensure economising through synergy effects.

The demographic state of the world described in this paper, its changes in the future decades, and the efforts undertaken to link population trends to development policy responses and measures, must be placed within the framework of ecological flow balances.

1. On the Demographic State of the World

A first overview is based on data compiled for the year 2002.[3] If the demographic state were to be related to just one specific year, we would receive an instantaneous photograph that does not illustrate the essentials. The population figure of one year, interlinked with the past and the future, is part of a dynamic process. Thus the perpetual question concerns the growth (or shrinkage) factors concealed in a population figure, as well as their strength of impact as regards their altering the population in a specific direction within one generation or, say, until the middle of the new century. (cp. Table 1).

In 2002, humanity comprised 6.2 billion, with 1.2 billion people living in ("mature") industrialised nations, 5 billion – in developing countries. The per capita gross social product, a reference criterion with a fairly inaccurate indicator function, amounts to US $ 22,000 for the former and US $ 3,500 for developing countries. However, it is the demographic discrepancies that are of crucial interest, since, taken together, they show a social state, a stage of development. It is doubtful, perhaps even undesirable, for the "Third Word" to enter the industrial channels of the Europeans by trying to imitate their "modernisation" instead of advancing their *own development* on a regional, cultural and ecological basis.[4]

---

3 Figures provided by the Population Reference Bureau Inc., Washington D.C., German version: Stiftung Weltbevölkerung *(World Population Foundation)*, Berlin, DSW data report 2002.
4 Peter Atteslander (Ed.), Kulturelle Eigenentwicklung – Perspektiven einer neuen Entwicklungspolitik. Frankfurt/New York (Campus), 1993.

Table 1: World population in 2002

	Population (mill.).	Birth figure in ‰	Death figure in ‰	Annual increase in %	Projection 2025 (mill.)	Projection 2050 (mill.)
World	6215	21	9	1.3	7859	9104
Industrialised nations	1197	11	10	0.1	1249	1231
Europe	728	10	11	-0.1	718	651
North America	319	14	9	0.6	382	450
Developing countries	5018	24	8	1.6	6610	7873
Latin America	531	23	6	1.7	697	815
Africa	840	38	14	2.4	1281	1845
Asia (excl. China)	2485	24	8	1.6	298	404
Japan	127.4	9	8	0.2	121.1	100.6
PR of China	1280.7	13	6	0.7	1454.7	1393.6
India	1049.5	26	6	1.7	1363.0	1628.0
Oceania (incl. Australia and New Zealand)	32	18	7	1.0	23.2	25.0

Source: 2002 World Population Data Sheet (Ed. Population Reference Bureau, Inc.), Washington D.C.

Taking a closer look at differences on a global scale, developing regions show a birth rate more than double that of industrialised nations. An annual percentage increase of 1.9% (not including China) is high. It is important to know that some 4% is the maximum that individual populations can achieve in their "explosion phase". This rate of increase means that the world population grows by a further 80 million each year. This figure concerns birth surpluses, whose decrease shows a gradually declining rate of increase. Declines of births began two decades ago, so projections must take this fact into account. They arrive at 7.9 billion for 2025, and 9.1 billion for 2050. Among developing countries, Africa is conspicuous for its vigorous growth, being expected to grow from its 840 millions in 2002 by one billion until 2050. Civil wars and AIDS epidemic do not seem yet to have any demographic impact on West Africa's 2.7% annual growth, although this is anticipated for the future. Another amazing development is the barely reduced growth of the Indian population, which will surpass the Chinese in twenty years; and finally the population of China that is curbing its growth with a strict one-child policy and does not intend to exceed its 1.4 billion. Population decline is already making itself felt in Europe. Until 2050, today's 726 million may have decreased by up to 100 million.

Therefore it is important to look behind the population figures to see the movement that influences them, as well as the speed and intensity of their change.

Fig. 1: World population (The 2000 Revision)

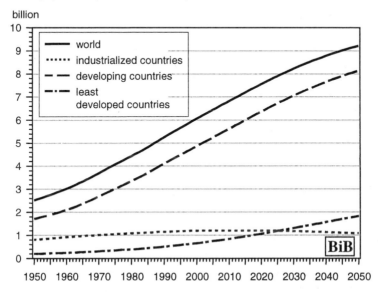

Source: United Nations Population Division (Editor), World Population Prospects: The 2000 Revision; cp. Wöhlcke/Höhn/Schmid (2003) and Bundesinstitut für Bevölkerungsforschung (BIB)[5]

Apart from the absolute population figure, the annual *rate of increase* in percent must be taken into consideration. It results from the increase quants of birth and immigration, minus the decrease caused by deaths and migration. For many regions of the world outside of Europe, where the surplus of births is the most important source of growth by far, the mainly politically influenced migrations are left aside; here, Europe is treated the same way, so as to make possible and illustrate the purely demographic comparison, i.e., the difference between birth and death rates.

---

5 Manfred Wöhlcke/Charlotte Höhn/Susanne Schmid: Demographische Entwicklung in und um Europa – politisch relevante Konsequenzen. (Nomos Verlag) 2003 (currently printing).

The *birth figure* (number of births in one year per one thousand of the population) is an important indicator for the state of development of a society, as well as the analogously calculated death figure.[6] As regards global development, one rule of thumb is that low birth rates indicate a high level of modernisation of the society, while somewhat higher rates suggest a modernisation potential that is not yet fully utilised. On this basis, the rough classification of industrialised and developing countries can be shown in the light of demographic indicators. Distinctly higher birth figures from 20 upwards (the Palestinian Gaza Strip is supposed to have the highest with about 50) indicate higher shares of agriculture and lines of business that require children as a family resource as well as jobs, such as has not been the case in Western Europe for a hundred years.

The *death figures*, too, suggest a specific stage of development, in particular infant mortality (deaths within one year of birth), which, in Western Europe, is only five cases in one thousand births, but between 100 and 150 in Black Africa. The death figures of industrialised countries and developing regions are not as far apart as the birth figures, a fact caused by the different age distribution. The age pyramid there is young, i.e., wide, with one half made up of young people under 20. If Latin America and Asia have a death figure of 6 and 8 (as against 11 in Europe), this only means that the highest share falls to the age groups with the least number of deaths: a high share of young as against a low share of an aged population over 60 with some 5%. This is the reason for high birth surpluses in the developing countries that can reach a net increase of up to 4% in individual regions (e.g., Algeria and Mexico in the sixties of last century). Today, the highest increase rates are found in Central Africa with 2.9%.

Europe's profit and loss account, calculated by births and deaths, shows in most cases an excess of deaths over births entailing a minus increase rate of currently -0.1%. Taken together, the industrialised countries – including European overseas offshoots like North America – still manage a minimal growth of 0.1%.

2. Comparing Age Structures

The growth dynamism governing the world population can be inferred from the comparison between "younger" and "older", i.e., modern population pyramids. The age structures of the "young" populations in the developing continents rest on a wide youth basis. This means that, only 20 years later, each of these strong

---

6 In journalese: "birth rate" and "death rate"; however, we are not talking about per cent but about "per mill" that are called "figure" rather than "rate". Only the increase, expressed in per cent, is correctly called "rate".

age groups will reach marrying age and strengthen the youth basis in its turn. Even if the birth rates show a slow decline in some regions, as reported for the last 20 years, the inherent logic of the age pyramid will yield so many parent couples that the rising number of marriages will make up for the decreased number of children, with little change of the annual rate of increase. The efforts of the global community aimed at reducing mortality also have an initially increasing effect, if only because more people will survive who will then have families, and through the decline of child mortality. In order to free the globe from its additional demographic burden – in view of its more than sufficient burdens from social, economic and political liabilities – it has become a central goal of world policy to bring the average increase rate of the developing continents to zero (2002: Africa 2.5%, Asia 1.5%, Latin America 1.8%).

The annual rate of increase merely shows the intensity of the upward or downward movement of a total population. We can take a look at one continent or one specific country. If a policy aimed at curbing population growth is to be conceived, we need to look at the inner life of populations, i.e., their growth dynamism. In this context, the development of births in families has a key position.

In Africa, the average number of children is currently 5.2, in Latin America – 2.7; however, we need to make a further distinction between cultural and climatic regions in continents to avoid using false averages as a policy basis. Arab West Asia, the Middle East, must contemplate the Arab north of Africa; Southern Central Asia with India is a separate region, while Eastern Asia seems to be culturally heterogeneous: apart from the neo-Confucian cultures (PR of China, Korea) it comprises the Islamic Indonesia, the Philippines with their lasting Hispano-Catholic imprint, and the Indochinese ASEAN states of Thailand and Vietnam with the most vigorous development efforts.

3. Projections

Several renowned organisations compile forecasts on the world population as a whole, as well as on continents and individual countries.[7] Even though popula-

---

7 In the UN system, it is the "Demographic Division" and the World Bank group: United Nations, World Population Prospects. The 200. Revision, New York, 200...; International Institute for Applied System Analysis/IIASA, A-Laxenburg: Wolfgang Lutz, Population – Development – Environment. Understanding their Interactions in Mauritius. Berlin 1994. idem: The Future Population of the World. What can we assume today? London 1994., Wolfgang Lutz, Warren Sanderson, Sergei Scherbov, The end of world population growth. In: Nature 412: 543-545; The World Conservation Union (IUCN): Gayl D. Ness/, Meghan V. Golay, Population and Strategies

tion processes like birth frequency and death probabilities of individual age groups reflect the social and economic circumstances of the respective world region, they interact and result in a specific annual rate of increase; it is the one-year surplus of births in per cent. For the world population of 6.2 billlion in 2002, a 1.3% rate of increase implies an annual increase of 80 million people. The world community is aware of the fact that in view of the finiteness of the globe, the closed nature of the biosphere, and despite the possibility of mobilising renewable resources, this human growth cannot go on forever. It is also well known that 90% of this increase is caused by the birth surpluses in the developing continents of Africa, Asia and Latin America. This is where four out of five children are born. India has the currently highest share of newborns.

Key variable: "Generation replacement"

The basic requirement for a standstill of the population growth is for families to have two surviving children (an average of 2.2 in order to compensate possible mortality). If parents with two children replace only themselves, this will be the basis for a gradual slimming of the wide population pyramid and an approximation to the form typical of Western countries (see fig. 1). However, even in the most favourable case this process will last a generation; it would require a change of socio-economic circumstances, so that having few children will be desirable and sufficient; the predominance of agriculture and small tenantry, with a traditionally high demand of children, must be reduced in favour of urban forms of gainful employment and services requiring vocational training for the children. Higher expenses, better chances of survival and the prospect of being able to support parents within an improved economic environment are the main motives for having fewer children.

However, the situation in the developing regions is not generally favourable enough for the motive of limitation of offspring to become effective everywhere with similar intensity and at the same time. The time in between has become an important projection tool. In view of the fact that strong young age groups become equally strong marrying age groups, the question is how long it will take for the youth basis to shrink slowly, for parents to decide to embark on the two-child system, until the youth age groups will not be much wider than the parent

---

for National Sustainable Development. London 1997, the private organisation Population Reference Bureau (PRB), Washington, D. C. Its annually compiled "Data Sheet" may not be highly exact, but is available worldwide. In Table 1, PRB data are being used.

age groups, until the Egyptian-type age pyramid gradually assumes a bell form, with the age groups lying on top of each other with approximately the same size. It remains to wait for (a) the turning point for changing the number of children per family to 2.2, and (b) the growth potential still inherent in the age pyramid to "outgrow" itself. Up to the point of growth stagnation, populations can continue to grow for the duration of one generation, increasing their population volume by one third. The growth inherent in the age pyramid that unfolds on the way to stagnation is called *"demographic momentum"*. It will determine the final size of the world population at the end of the 21st century.

The reports and information about an annual decline of net increases, of the average number of children in families worldwide, are still no reason to feel reassured, to lean back comfortably, assuming that "the population bomb has been disarmed". Even if demographic indicators show a decrease of the population pressure, there is still the growth momentum to be considered.

Table 2: Estimated and projected average number of children per woman in different regions of the world (1995-200 and 2045-2050)

Region	Total fertility (average number of children per woman)			
			2045 – 2050	
	1995-2000	Low	Medium	High
World	2.82	1.68	2.15	2.62
Industrialised countries	1.57	1.52	1.92	2.33
Developing countries	3.10	1.70	2.17	2.65
"Poorest countries"	5.47	2.02	2.51	3.02
Africa	5.27	1.91	2.39	2.88
Asia	2.70	1.60	2.08	2.56
Latin America and Caribbean	2.69	1.60	2.10	2.59
Europe	1.41	1.41	1.81	2.20
North America	2.00	1.68	2.08	2.48
Oceania	2.41	1.61	2.06	2.50

Source: United Nations Population Division, World Population Prospects: The 2000 Revision

On the global average, the number of children per woman amounts to almost 3 children (2.82), and this implies that the parent generation would increase by one and a half. A striking discrepancy can be observed between industrialised nations, with 1.57 distinctly below the birth target needed for stock maintenance, and the developing continents with 3.1. Africa has the highest birth figures (5.27) and also the largest share of the "poorest countries" (5.47); Asian averages are dominated by the population mass of China. Its strict one-child policy has the effect of a downward distortion of Asia's birth average. This is why China is frequently left out of continental statistics and listed separately (2002: 1.8 children).

Fig. 2: UN world population projections – 2002 Revision (three variants)

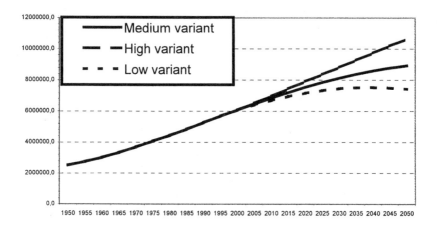

Source: United Nations Population Division (Hrsg.), World Population Prospects: The 2000 Revision.

If growth were to continue in the same way as until 2000, the world population would have arrived at 13 billion by the end of the projection period, having more than doubled within 50 years. Since, however, declining increase rates must be considered on the basis of decreasing birth rates, i.e., their intensity must be estimated, projections will also have to be much lower. The projections for the world population until 2050 show three lines. A lower variant reckons that the highest-population nations will reach the "reproduction level" of a 2.2 children average as soon as 2020, which would imply a fairly improbable low population figure below 8 billion. A high variant expects this only by the year 2050, which would imply a longer unfolding time for the demographic growth momentum. At the UN Population Division, a medium variant is being used, which projects 9.3 billion people for the year 2050. The ultimately stabilised size of the world population, which will probably not be reached by the end of this century because of Africa, the problematic latecomer, will most likely amount to slightly above 10 billion.

4. Demographic Transitions in the Development Process

Rapidly growing populations must undergo a profound social change in order to be able to reduce high birth surpluses. The population processes are interwoven with all areas involved in the development and modernisation process: the decline of the increase rates entails a change of age structures, which get "older", transforming their "traditional" form, suitable for a rural culture, to a "modern" culture that corresponds to the requirements of an industrial mode of work and living: few children in a mobile small family, deaths are the exception, a well-developed health system ensures a rising life expectancy.

During the development process, populations pass through growth stages. They grow by the surplus of newborns over deaths (at the different stages of life). This surplus increases from the moment when investments, technical and medical measures are used to reduce death rates, in particular in childhood. Europe has finished this growth phase with its subsequent decline of births, and has ended up by even surpassing it. It has been living with death surpluses for years. Third-world countries are still at earlier stages of their path towards demographic solution, their "transition"; it is easier to reduce deaths than to persuade millions of young couples that they will not need the accustomed number of births to ensure the desired number of surviving children.

Europe and North America have reached the final phase of demographic transition with only 1.4 children per woman. This implies that figures in the modern world are already one third below the parental generation size with 2.1 children per woman. East Asia is next, well-known for its efforts to solve its population problem like the Europeans, with the aid of economic progress. Yet West and Middle Asia, from the Middle East via the Indian subcontinent to Indonesia, is in the growth phase with low death figures and high birth rates that decline very sluggishly; the same applies to tropical Latin America. Africa is in an intensified growth phase, with the largest gap between birth level and mortality. Before the end of the next century, it is not likely to reach the stagnation phase, that final phase of demographic transition when births and deaths converge at a low frequency level, with mere population replacement but no population growth.

This shows that all populations must go through a stress phase that can be understood fully only by considering that mastering it requires the development and emergence of new ways of providing food, technologies, resources and industrial forms of organisation.

5. Population-related Global Problems

5.1 Age structure

The age structure alone suggests the problems lying in wait for most developing countries. The strong young age groups are pushing inexorably into the poorly developed health and food sector, into schools and educational facilities, into the housing and job market. Moreover, since they will be the enlarged marrying age groups 15 to 20 years later, this will lead to a structurally based "children's children" effect that compensates and surpasses even a moderate decline of births. In combination with the generally declining mortality thanks to modern medical imports, this has the effect of maintaining the "growth gap".

Above all, the schooling of the swelling young age groups poses enormous problems. Whereas industrialised nations worry about the burden of providing for growing old age groups, developing countries must deal with the *youth burden*, i.e., the ratio of working population to young age groups. Almost half the population is under 15 (compared to a mere 15% in Germany). The People's Republic of China has coped with this problem through organisational methods, while in the Latin American, and even more the African continent, literacy and education efforts stay far behind the rapid population growth. The disproportion between existing educational facilities and the size of the school age groups can improve only if educational investments and family planning are adequately combined. In the case of Black Africa, this will probably not happen before the middle of the next century.

In the developing countries, the population that is capable of gainful employment will grow from some 1.76 billion today to more than 3.1 billion people in 2025. Every year, 38 million new jobs will be needed, not counting those that would be necessary for eliminating the existing unemployment, which is at 40% in many developing countries.[8] The rural surplus population can hardly be absorbed by the weak employment capacities of industry and trade. The service sector in the cities is blown up artificially, serving as a cover for concealed unemployment. As a result of low equity capital formation, the worldwide decline of raw material prices together with increasing prices for industrial goods (terms-of-trade effect) and economic problems, also in the industrialised nations (protectionism), the Third World labour markets do not expand fast enough. Help is envisaged only through a "gentle" industrialisation of the rural regions, which could stop the alarming urbanisation.

---

8 D. E. Bloom/A. Brender, Labor and the Emerging World Economy, Population Bulletin (Pop. Reference Bureau, Washington D.C., Vol. 48, 2. October 1993).

The debt crisis of the Third World is based not only on the desperate efforts of many governments to obtain in the international capital market the financial resources required for providing produce and services for their rapidly growing populations. The frequently unproductive jobs and the grotesquely blown up administrative machinery are the local versions of the electoral democracy implanted by the West. This problem is aggravated by the transfer of new *technologies* which, having been developed in the West, aim at the saving of labour.

The supply burden inherent in every age structure (age pyramid) tangles the population and development problems. The age structure is the result of preceding fertility, mortality and migrations in all age groups. The central defining quantity is made up by the births, which put the population pyramids of the developing countries on a wide youth base. They contain the amount of future producers of working age, indicating that the demand for educational expenses and capital is rising super-proportionally. Even where economic progress was observed and improved market opportunities for products were finally converted into an increased deployment of labour, the youth burden makes itself felt, again weakening the initiated positive tendency. We speak about demographic costs caused exclusively by the large offspring. Beyond just earning their living, young people must be trained for rising demands to the quality of labour, i.e., the educational expenses for them are increasing faster than the consumption of the remaining population groups. This means that the educational expenditure for children can be reduced only relatively via an economic boom, or absolutely through a limitation of births. In countries where crowds of children are sorting garbage, eking out a bare existence in the streets by doing menial services on the basis of tips, neither the one nor the other seems to work. It can be said that the youth burden limits economic progress, because supply and educational costs, provided they can be raised at all, swallow up funds earmarked for development purposes. The birth reduction policy cannot be seen under the sign of "fewer people" but as a means of relief on the development path.

5.2 Ageing – Not Only in the "Old World"

The population growth in the developing continents forms a stark contrast to the demographic state of Europe. In Europe, the ancient human dream of a long life has come true. Yet solidary communities cannot remain idle: ageing has made itself felt. Birth decline has long developed into a "deficitary birth output", since with 1.3 children per woman in Central and Southern Europe rates have fallen to one third below generation replacement, and old age groups are increasing their shares. Besides, generous health services ensure a growing life expectancy, in particular in the ranks aged between 80 and 100 years.[9] In Western Europe, it was 74 for men and 80 for women in the year 1999: between 2040 and 2050, today's industrialised countries will have an average life expectancy of 77.7 for men and 83.8 for women.[10]

Table 3: Average life expectancy in years (from 1950 – 2050)

	World	Industrialised countries	Developing countries	"Poorest countries"
1950-1955:				
Men	45.1	63.9	40.1	34.9
Women	47.8	69.0	41.8	36.2
1990-1995:				
Men	62.2	70.4	60.6	48.7
Women	62.5	78.0	63.7	50.8
2020-2025:				
Men	69.7	75.1	68.8	62.0
Women	74.5	81.6	72.9	64.7
2025-2050:				
Men	73,8	77,7	73,2	69,3
Women	78,8	83,8	77,8	72,9

Source: United Nations, World Population Prospects. The 1996 Revision. New York 1996

The problems resulting from the changes of the age structure occur differently and with varying intensity in the different parts of the world. In Europe, they exert funding pressure on the social security systems and challenge the hitherto peaceful inter-generational relationship. In the developing continents, the demographic burden is solely on the part of the adolescents. They must be carried along on the development path until they become human capital and help to bear the burden of development.

---

9 Josef Schmid, Population Ageing: Dynamics, and Social and Economic Implications at Family, Community and Societal Levels. Paper at the meeting of UN/ECE (Geneva), Budapest, 7.-9. December 1998. (CES/PAU/1998/6; GE 98-32457)
   Gérard Calot/Jean-Paul Sardon, Les Facteurs du vieillissement démographique. In: POPULATION, 54 (3), 1999, pp. 509-552.
   Wolfgang Lutz, Brian C. O'Neill, Sergei Scherbov, Europe's Population at a Turning Point. In: Science, Vol 299, 28.03.2003, pp. 1991-1992.
10 Data Sheet 1997; United Nations, World Population Prospects. The 1996 Revision, New York, United Nation 1996; Andreas Heigl/Ralf Mai, Demographische Alterung in den Regionen der EU. In: Zeitschrift für Bevölkerungswissenschaft 3/1998, pp. 293-317; Wolfgang Lutz, Brian C. O'Neill, Sergei Scherbov, Europe's Population at a turning point. In: Science 299: 1991 – 1992, 28 March 2003.

From the year 2015 onwards, the former children of the sixties' baby boom will begin to retire. Their inflow into the pension age will continue until 2030; they will find themselves confronted with low-birth active age groups of the ensuing sharp drop in births related to the introduction of oral contraceptives, which has continued since the seventies until this day. By the beginning of the century, the ratio of the under-twenties to the over-sixties will have been reversed. By 2025, older people will make up some 30 per cent of the total population.

This has long been the governing trend for the white US population, the populations of Japan, Singapore, Korea, and above all the population mass of the PR of China, provided that it maintains the strict one-child policy, using population policy to create the "top heavy" structure which the Europeans have generated with their lifestyle.

5.3 Urbanisation

Global problems have increased inasmuch as sciences have begun to deal with them, defining each of them as a separate area with specific competences. "Food security" falls within the range of agricultural policy, agricultural research and population growth.[11] The World Health Organisation (Geneva) supervises clinical pictures and causes of death, in particular centres of epidemics and AIDS infection rates. The International Labour Office (ILO, Geneva) calculates occupation chances in view of large working-age groups in developing countries, as well as with regard to unemployment in high-technology societies. There are several competences for migration movements, in particular working migration, illegal border crossings, human smuggling and organised criminality.

In the urbanisation phenomenon, all demographic and social problems of the world population seem to converge: the inner population pressure caused by birth surpluses in rural regions intensifies the pressure on the cities. In the Third World, this is no "urbanisation" after the European model, where a rural-exodus population mutated into urban citizens within one generation, with their new way of life and intellectual flexibility. Rather, huge numbers of makeshift homes are shooting up in enormous haste for people who live on the vague hope to have taken the right step for themselves and their children. At the beginning of the 21st century, for the first time in human history, more people are living in cities than in rural regions.

---

11 Klaus M. Leisinger, Karin M. Schmidt, Rajul Pandya-Lorch, Six Billion and Counting-Population and Food Security in the 21st Century (The John Hopkins University Press) 2002.

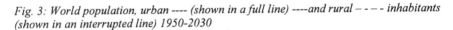

Fig. 3: World population, urban ---- (shown in a full line) ----and rural – - - - inhabitants (shown in an interrupted line) 1950-2030

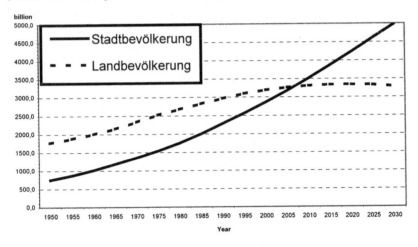

Source: *United Nations, Population Division, World Urbanization Prospects: The 2001 Revision*

From 1995 to 2030, urban populations will double, growing from 2.6 billion to 4.9, 4 billion of which will live in developing regions. By 2020, over half the populations of Africa and Asia will live in cities; three fourths of Latin Americans are living there already.[12] The largest national populations, China and India, are leading the push into the cities. Whereas two out of three Chinese are currently living in rural areas, over half the Chinese population will be urbanised by 2030. The same applies to India, Japan, Russia and the United States. This is an increased and unstoppable trend towards agglomeration, which will cause a decline of the rural population from 2015 onwards. Impoverished rural populations will soon end up in slums, shanty towns or "favelas", where it will remain in the lowest menial services, conditions of exploitation, beggary, petty criminality or prostitution. There is an overwhelming feeling that a social betterment or ascent of the children into normal jobs can only be achieved in urban regions. For this goal, people put up with evils from which they were spared in their rural regions. For example, the share of undernourished children in the cities has risen sharply.

---

12 J. L. Garrett, Overview. In: J. L. Garrett, M. T. Ruel (eds.), Achieving Urban Food and Nutrition Security in the Developing World. 2020 Focus 3 (Washington, D.C.:IFPRI, 2000).

The sanitary conditions of crowded masses lead to epidemics, drug abuse and HIV infections. Impoverished urban inhabitants cannot provide for themselves anymore. Even insufficient nourishment requires "money" and has its "price". *"Thus food security in urban areas is inextricably linked to income security."* [13]

As long as governments seem to care only about cities, allotting them the main part of investments, people will continue to leave their rural regions.

The mega-metropolises of 12 to 27 million inhabitants to be expected for the year 2015 are listed in Table 4. Only three of them (Tokyo, New York, Los Angeles) lie in the modern world.

Table 4: The largest conurbations with estimated growth until 2015 (population in 1000)

Country	City	1950	1975	2000	2015
Japan	Tokyo	6 920	19 771	26 444	27 190
Bangladesh	Dhaka	417	2 173	12 519	22 766
India	Mumbai (Bombay)	2 981	7 347	16 086	22 577
Brazil	Sao Paulo	2 528	10 333	17 962	21 229
India	Delhi	1 391	4 426	12 441	20 884
Mexico	Mexico City	2 883	10 691	18 066	20 434
United States	New York	12 339	15 880	16 732	17 944
Indonesia	Jakarta	1 452	4 814	11 018	17 268
India	Calcutta	4 446	7 888	13 058	16 747
Pakistan	Karachi	1 028	3 990	10 032	16 197
Nigeria	Lagos	288	1 890	8 665	15 966
United States	Los Angeles	4 046	8 926	13 213	14 494
China	Shanghai	5 333	11 443	12 887	13 598
Argentina	Buenos Aires	5 042	9 144	12 024	13 185
Philippines	Metro Manila	1 544	5 000	9 950	12 597

Source: United Nations Population Division, World Urbanization Prospects: The 2001 Revision.

The absolute growth from 1950 onwards illustrates the incredible intensity of the agglomeration process, which can only be grasped with the help of satellite photographs. Mega-cities with over 11 million inhabitants are subsequently found on this list: Peking, Rio de Janeiro, Cairo and Istanbul.

The economic and social policy problems entailed by the administration of such agglomerations are beyond the imagination of Europeans used to orderly

---

13 Klaus M. Leisinger, ibid. p. 12.

conditions. Environmental problems reach dramatic proportions. All third-world cities have poisoned areas and waters whose vicinity should be avoided. Energy supply, e.g., electric power, is eccentric: public lines are tapped by the thousands. The primary concern of ecologists is about the location of these conurbations, since practically all of them are situated at the sea. This implies an enormous pollution of the seas, with a negative impact on still existing regeneration areas.

5.4 Environmental Problems

System and prognosis research experts have developed impact and flow charts that greatly facilitate the demonstration of environmental crises. We will give you an example of the population-environment relationship, developed within the framework of a demographic educational programme (Population Education).[14]

Inserted between *population* and *environment* are the agents of culture in the narrower sense: organisation and social structure (defined here as *society, culture, consumption, trade*). Technology in evolutionary systems is always *technology development*. Population is broken up into its components; on the one hand, into the population processes (*fertility* = level of births, *mortality,* and *migration*), on the other – into the structural categories of *size, distribution* and age *structure.* Environment is subdivided into renewable resources *(air, water, energy, land),* into the purely physical environment *(non-built, uninhabited)* and the "cultural landscape" wrested from "nature" as *built environment.* Governmental intervention and international interconnections complete the complex.

The relationship of population to resources, which used to be called "food scope", led to the question of the "*carrying capacity"* of the soil. Carrying body research occupied a generation of cultural geographers and agro-economists.[15] In the meantime, research has turned to the biosphere as the reference value of human development.

---

14 John I. Clark, Education, Population, Environment and Sustainable Development. In: International Review of Education (Special Issue: Population Education), UNESCO Institute for Education, Hamburg; Vol. 39, Nos. 1-2, March 1993, p. 56; see also: Josef Schmid, Bevölkerung – Umwelt – Entwicklung. Eine Humanökologische Perspektive. Opladen 1994.

15 Wolfgang Kuls, Probleme der Bevölkerungsgeographie. Darmstadt 1978. Jürgen Bähr, Chr. Jentsch, W. Kuls (Ed.), Bevölkerungsgeographie. (Lehrbuch der Allgemeinen Geographie, Vol. 9), Berlin-New York 1992, p. 117 ff.

Fig. 4: Factor model of the population-environment relationship

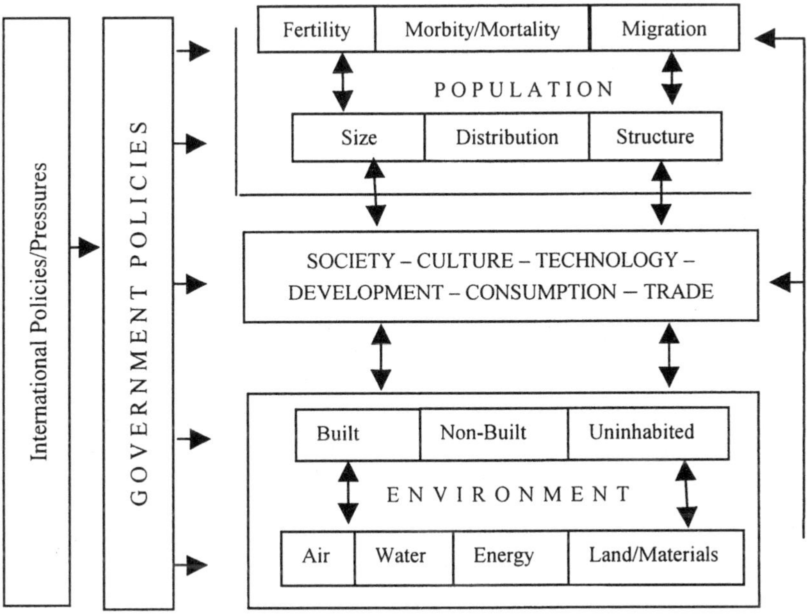
International Policies/Pressures
GOVERNMENT POLICIES
Fertility
Morbity/Mortality
Migration
POPULATION
Size
Distribution
Structure
SOCIETY – CULTURE – TECHNOLOGY –
DEVELOPMENT – CONSUMPTION – TRADE
Built
Non-Built
Uninhabited
ENVIRONMENT
Air
Water
Energy
Land/Materials

Source: Clark (1993), p. 56

Environment is what remains as a result of the adjustment struggle between population and resources. The population-environment utilisation relationship rests on the following three components:

• Lifestyles, income and social organisation, which combine to determine the level of consumption

• The technologies in use, which determine to which extent the social activities harm or preserve the environment; they are definitely related to the consumption level. Components 1. and 2. can statistically be related to the individual person.

• Population: this is the multiplier of the average influence per person that yields the total impact.

Extreme *inequalities* in the distribution of the land that force the poor to make the maximum use of small and marginal plots of land, become an additional factor if they prevent the introduction of environmentally favourable technologies, thus intensifying the harmful impact on the environment.

The highest industrialised nations have the highest share in the consumption of resources. With 25% of the world population, they consume 75% of the total world energy, 79% of all fuels, 65% of all timber products, and 72% of the total steel production. At the same time, they cause three fourths of the total carbon dioxide emissions, which are responsible for half the greenhouse effect in the atmosphere. This is only counterbalanced by the fact that they are the main producers of the world economy, its surpluses and financial resources, on which the developing countries have long come to depend.

The "poorest of the poor", living in the developing countries, are the ones that have the most pressing need of the blessings of development, but are compelled to destroy their own resource basis from economic need and from a lack of alternatives. They cause part of the environmental destruction in the southern hemisphere. Their growth rates indicate a doubling of their populations approximately within the next 25 years. Projects for lessening or even halving their misery do not appear to pay adequate attention to this fact. Situated between them are the better-off developing countries and threshold countries, with no less environmental destruction but much better chances of using technological process to link up with the policy of closed circuits.

The losses of renewable and unrenewable resources will progress rapidly unless technology and investments can stop this process. Greenhouse effect and ozone hole have already shifted the problems into the atmosphere and biosphere. Recent findings show that the greatest menace does not come from depletable resources but from the negative impact on the renewability of the sustainable resources that are of direct importance for life, i.e., soil, air, water and biodiversity.

The soil question is being discussed in relation to the problem of soil erosion, the advance of the deserts and the dramatic decline of the forests, in particular of tropical forests.

6. A different Demographic Transition for developing Countries as a contribution to Global Sustainability

In a certain way, the theory of demographic transition (see chapter 4) is an anchor for science and politics that population science drops when asked to comment on past and future trends.[16]

---

16 Josef Schmid, Bevölkerung und soziale Entwicklung – Der demographische Über-

The demographic transition is a hoard of experience of European social history and modernisation that was frequently seen as a suitable way for non-Europeans as well: certain development steps reduce mortality, intensifying so that increasing prosperity gives rise to the modern urban alternatives to the old rural family life. The surviving children are considered a burden, since expenses for children that are regarded as too high might endanger a comfortable lifestyle (leisure time, weekends). This explanation is derived from the Anglo-Saxon narrow-minded theory of utilitarism. In Germany, this process was soon cast into the affluence theory of birth decline. The Malthusian model was finally dethroned, since it had been found that – contrary to the Malthusian theory – the increase of the family income does not lead to a growing number of children, but just the opposite: the subjective *credo* that the individual strives for an optimum of success, even in partnerships. As the blessings of the industrial age increase, former sufferings from hunger are being replaced by the trouble of being spoilt for choice, by rivalling pleasures. The rising standards entail rising restlessness and awareness of crises, creating the modern unsteady social character. Family life and children are increasingly contradictory to the *zeitgeist*. Obviously the benefit previously expected from numerous offspring can now be realised with 1 – 3 children. Interest in family planning grew, finally closing the gap between birth and mortality levels that had opened during the transition stage.

European-type demographic transition requires three generations. This theory of population development, a counterpart to the theory of modernisation, has remained to this day one of the few accepted development laws. Certain nervousness has seized the discipline, since it needs to part with a familiar conception or will have to extend it beyond recognition in order to take the future state of the world into account. Two facts will enforce a change of paradigms:

First of all, the fertility of the Third World will decrease without finding support in real development and enhanced industrialisation. China might demonstrate the case in an exemplary fashion.

Second, the developing countries will have to manage their demographic transition in a different way from today's industrialised nations. Nathan Keyfitz, former director of IIASA, Laxenburg, puts it like this, that there is a deep connection between energy consumption and income, and if this correlation still ex-

gang als soziologische und politische Konzeption. Schriftenreihe des Bundesinstituts für Bevölkerungsforschung, Boldt-Verlag, Boppard/ Rhein, 1984
John Caldwell, Toward a Restatement of Demographic Transition Theory. In: Population and Development Review, Vol. 2, No. 2/3, 1978, pp. 326-366.

ists at present, developing countries will never have the chance to have the same level of income like industrialized countries.[17]

This implies that the demographic transition in the Third World will neither come about at the easy pace of the North and West European model, nor will it take the same route. If a development is to be envisaged for the second half of the 21st century, the Third World will be forced to undergo the following changes:

The Third World cannot take a hundred years to complete its demographic transition, because populations ten times bigger are awaiting development, with growth rates Europe has never experienced.

The externally initiated reduction of mortality must entail a faster decline of births, since the growth gap at an annual 2% net increase results in such large birth surpluses that they cannot be coped with for a longer period of time.

This means that the developing countries need external help for reducing their birth rates just as for the reduction of mortality. Reduction of births cannot wait for the general development process to take its course. It will appear like a *reversal of the European way*: demographic modernisation must have been set in motion and have the effect of facilitating the development process that is still lagging behind.

The Third World will not be able to pursue its development with the same degree of exploitation of nature and waste of energy as the northern hemisphere. Here, too, a reversal will be required. The ecological and climatic conditions require the premature (!) introduction of intelligent, partly also of traditional, low-energy systems. Another way for ensuring a tolerable existence for at least 8 billion people in the South cannot possibly be conceived.

To sum up it can be said that global sustainability will get closer to its goal if the Third World develops according to a different priority scheme: birth rates must follow faster after low mortality, reduce birth surpluses faster and fit smaller-sized age groups into low-energy systems with environment-renewing technology. This requires investments and educational expenses of a scope that cannot be raised on their own by populations with a 50%-share of children and adolescents.

---

17 Nathan Keyfitz, Completing the Worldwide Demographic Transition: The Relevance of Past Experience. Ambio 21 (1), 1992, pp. 26-30.

Fig. 5 Young- (0-20) and old age groups (60+) in Europe

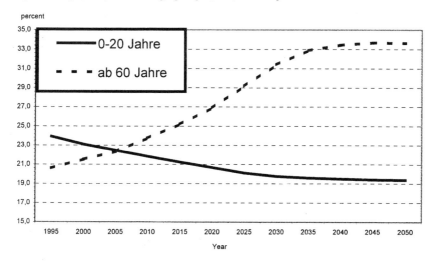

Source: Eurostat

Climate Change: Mitigation and Adaptation

Mahendra Shah

⇒ Alternative energy ressources are the solution for the future.
⇒ At least 40 developing countries may lose up to quarter average of their cereal-production potential due to climate change in the 2080s.

1. Introduction

Combating climate change and variability is vital to the pursuit of sustainable development. Equally, the pursuit of sustainable development is integral to lasting
climate change mitigation and adaptation. The climate change issue is global,
long term and is a complex interaction between climatic, social, environmental,
economic, technological, institutional, and political processes. It has significant
international and inter-generational implications.

Climate change and variability affects the thermal regimes and through it, the
hydrological regimes. This, in turn, affects the structure and functionality of ecosystems and human livelihoods. Reports of increased climate variability and extreme events are becoming more and more frequent. In the absence of mitigation
and response capacities, losses resulting from infrastructural and economic damage will only escalate. The world will also face social turmoil and loss of life.
This burden will fall on the poorest nations and their people. It is only in poor
countries that drought turns to famine, often resulting in population displacement,
suffering, and loss of life. The social and economic costs of such occurrences
may undo, in just a day or a month, the achievements of years of development efforts.

The central challenge of sustainable development in the 21st century is to
meet the needs of the present generation without sacrificing the livelihoods of
future generations. This cannot be achieved without systemic integrations of the
social, economic and environmental pillars of development, and, a global partnership for equity and sustainability.

We live in a world of disparities where a fifth of the population lives in degrading poverty on about 1% of the global income, while another fifth consumes
more then four-fifths of the world's income. In 1990, developed countries with a
quarter of the world's population were responsible for over 70% of the cumulated
CO_2 and annual greenhouse gas emissions.

Vulnerable populations have only a limited capacity to protect themselves from the consequences of environmental change. They will bear the brunt of the impact of climate change, including land degradation and reduced productivity, biodiversity loss, an increased health risk and scarcity of water resources.

In the short term, policymakers will need to cope with an increased risk of frequent shocks to their economies, which will affect the welfare of their most vulnerable populations. Over the long term, they will need to cope with the effect of climate change on the underlying production structures of the economies. Those countries dependant on natural resources are most at risk.

2. Climate Change

During the 20th century, human influence on the functioning of the Earth's systems altered the global carbon pools. As a result, carbon dioxide concentrations in the atmosphere increased by over a quarter. Also, the concentration of methane, another important greenhouse gas, has doubled. There is now convincing scientific evidence that greenhouse gases have contributed to climate change.

Article 2 of the UN Framework Convention on Climate Change[1] establishes the "stabilization of greenhouse gas concentrations in the atmosphere at a level that would prevent dangerous anthropogenic interference with the climate system...stabilization should be achieved within a time-frame sufficient to allow ecosystems to adapt naturally to climate change, to ensure that food production is not threatened and to enable economic development to proceed in a sustainable manner".

A major finding of the IPCC Third Assessment Report[2] is that over the 20th century, the temperature has increased by 0.6 + or − 0.2 deg C. The IPPC assumed greenhouse gases as the primary driving force behind climate change, and based on the assumptions of future development paths, projections indicate that the temperature will rise by 1.4 to 5.8 deg C. A number of other interacting and feedback phenomena such as cloud cover, water vapour, aerosol, ocean currents, solar cycles and irradiance, etc also affect climate change and need to be considered in future climate change projections.

---

1 United Nations Framework Convention on Climate Change, UNCED, Rio de Janeiro, Brazil, 1992
2 IPCC, 2000, Summary for Policymakers, Emissions Scenarios, A Special Report of IPCC Working Group III, Intergovernmental Panel on Climate Change, ISBN: 92-9169-113-5.

In 1997, the international community negotiated the Kyoto Protocol[3], in which the developed nations agreed to limit their greenhouse gas emissions, relative to the levels emitted in 1990. Energy consumption is by far the largest emitter of greenhouse gases. There is a need not only to adopt energy conservation and efficiency measures, but also to increase the share of renewable energy. At the global level, the total energy consumption in 2000 amounted to 9958 Mtoe, comprising: 42.7% oil, 16.1% gas, 7.9% coal, 3.7% solar, wind, and geothermal, 13.8% combustible renewable energy and waste, and 15.8% hydro electricity. Adopting clean, renewable energy sources would contribute substantially to reducing greenhouse gas emissions and a concerted effort is required to develop and use clean renewable energy sources, particularly, solar, wind, hydro, wave power and tides, and geothermal. There is also scope to use biomass energy, thus reducing pressure on the demand for fossil fuels.

The Kyoto Protocol has also proposed economic instruments such as the 'Clean Development Mechanism' with certifiable emission reductions and land use and forestry changes dealing with carbon sinks. The progress in implementation has been difficult, as national governments have been politically constrained. Since the reductions entail modifying high fossil fuel consumption life-styles, consumers are as yet unprepared to make the necessary consumption changes. Reducing emissions is only part of the story. The current levels of greenhouse gas concentrations in the atmosphere imply that climate change in the next half-century is inevitable.

Responses to climate change can be of two types: adaptive measures to reduce adverse impacts and risks and to maximize the benefits and opportunities created by climate change; and mitigation measures to reduce anthropogenic contributions to climate change. Both adaptive actions and mitigation measures are necessary elements of a coherent and integrated response to climate change. If we are to stem global warming, we have no choice but to reduce the rapidly increasing emissions of greenhouse gases such as carbon dioxide. But in doing so, the contribution to and consequences of such emissions, as well as the different national development needs and priorities, have to be central in reaching economically efficient and environmentally effective agreements.

While many developed countries have assessed the impact of climate change on their own economies and natural resource environment, most developing countries have not done so. International negotiations are often constrained in an environment where one group of countries is well informed and another group,

---

3 Kyoto Protocol to the United Nations Framework Convention on Climate Change, Kyoto, Japan, 1997.

less so. A concerted worldwide effort is required to carry out national and regional assessments of potential impacts of climate change and variability.

Climate change has a long-time horizon, since its impact will emerge only over the next hundred years. However, effective response strategies have to be implemented in a timely fashion, so that they can be sustained over the coming decades. At the same time, response policies and measures are local whereas the consequences are global in nature. Both aspects, i.e., the long-time horizon and the global nature of the climate problem, together with the scientific uncertainties they present, pose special challenges for decision makers that have to balance potentially demanding action for averting global long-term risks against other, more immediate, human development demands.

Many factors contribute to vulnerability, including rapid population growth, hunger, poor health, low levels of education, gender inequality, a fragile and hazardous location, lack of access to resources and services, production and consumption patterns, international commodity and input prices, aid and investments, and knowledge and technological means.

Our knowledge of climate change cannot be perfect due to the many uncertainties that surround it. However, it is essential that we act in accordance with the "precautionary principle" to reduce the risks we face. Without this, it is likely that there will be a significant climate change, which will impact the life support capacity of ecosystems services, particularly with regard to land and water resources and biodiversity.

The past 50 years have given cause for worry; rapid land-cover changes, biotic fluxes, and the extinction of living species have been observed. The disturbing truth is that we do not even know the extent and the nature of the biodiversity that is being lost around the world – in our forests, in the oceans, and on land. The need for food for an increasing population is threatening natural resources, as people strive to get the most out of land already in production or push into virgin territory for agricultural land. The damage we are inflicting on the environment is increasingly evident: arable lands lost to erosion, salinity, desertification, and urban spread; disappearing forest and threats to biodiversity; and water scarcities.

Global biodiversity is under particular risk from climate change. Already hemmed in by habitat loss, pollution and over-exploitation, species and natural systems are now faced with the need to adapt to new regimes of temperature, precipitation and other climatic extremes. At the species level, those that are already critically endangered because of existing pressures are likely to be driven to extinction by the added stress of climate change. Migratory species will also be at risk since they require separate breeding, wintering and migrating habitats.

Freshwater resources are an essential component of the earth's hydrosphere and an indispensable part of all terrestrial ecosystems. Two-thirds of the world's population lives in areas that receive a quarter of the world's annual rainfall, while such sparsely populated areas as the Amazon Basin receive a disproportionate share. About 70% of the world's fresh water goes to agriculture, a figure that approaches 90% in countries that rely on extensive irrigation. At present, approximately 1.7 billion people, of a population of around six billion, are living in countries experiencing water stress. The UN Comprehensive Assessment of the Freshwater Resources of the World[4] estimates that by 2025, around five billion – of a total population of some eight billion – will be living in water-stressed countries. Climate change has the potential to alter these patterns of stress. Some parts of the world will receive more river runoff, while other parts will see a decrease in resource availability. Climate change will affect the availability and quality of freshwater resources and, through sea-level rise, will threaten low-lying coastal areas and small island ecosystems.

3. Climate Change and Food Production

Land and water resources, forest ecosystems, and biodiversity together with agricultural technology form the resource foundation for the functioning and sustainability of food production systems. Maintaining the fertility and multifunctionality of soils, preserving genetic diversity and adopting effective water resources management and protection measures are critical to enhancing agricultural production. By the same token, agricultural practices that include inefficient fertilizer and pesticide use, lack of land and water conservation measures and large-scale conversion of forest areas will result in irreversible damage to ecosystems and an accelerated rate of loss of production potentials. We cannot be complacent, not when the foundation of human survival, i.e., the need for food, is put at risk due to climate change.

Although at the global level, food production is sufficient to provide adequate nourishment for all, there are 84 countries in the world with a population of some 4 billion, of which about a fifth are chronically undernourished. By the 2080s, the total population of these currently food-insecure countries is projected to increase to some 7 billion, equivalent to 80% of the world's future population.

---

4 UN Comprehensive Assessment of the Freshwater Resources of the World, WMO, Geneva, Switzerland, 1997.

A recent IIASA study[5] commissioned by the United Nations for the Johannesburg 2002 WSSD, carried out a uniform and an integrated ecological-economic assessment of the impact climate change had on agriculture across all countries and regions of the world.

The country-level and regional results compare the ecological and economic impacts on agriculture, and particularly the food systems.

The results of this study indicate that while developed countries in the temperate zones substantially gain in agricultural production due to climate change, many developing countries in the tropics lose. Individual country results offer reason for concern. For example, some 40 least developing countries may lose up to a quarter average of their cereal-production potential due to climate change in the 2080s. In these countries, the average domestic per capita food production has declined by 10% in the last 20 years, in sharp contrast to an average increase of some 40% in the developing world. In these least developed countries, which account for 10% of the world's population and less than 1% of the world's income, the number of undernourished has doubled to some 250 million in the past two decades.

Sub-Saharan Africa has the highest percentage of undernourished – some 40% of the total population – and there has been little progress on reducing the hunger in the last three decades. The results of the IIASA study highlight the plight of many countries. For example, Sudan, Nigeria, Senegal, Mali, Burkina Faso, Somalia, Ethiopia, Zimbabwe, Chad, Sierra Leone, Angola, Mozambique, and Niger lose cereal-production potential for projected future climate change, even if one assumes a very optimistic future development path.

4. Climate Change: Fairness and Equity?

Global warming raises the issue of fairness. The total carbon dioxide emissions of developing countries, which account for more than four-fifths of the world's population, amount to less than a quarter of global emissions. Yet, it is many of these developing countries that will suffer substantially from the negative impact of climate change.

An extreme example is that of Mozambique, which is already at the mercy of annual, recurrent droughts or floods during the last 20 years. It is a country with 18 million people, of whom over 70% are undernourished, with a per capita deficit of a fifth of the minimum daily calorie requirement. This country produces 0.1

---

5 G Fischer, M M Shah and H van Velthuizen, Climate Change and Agricultural Vulnerability, WSSD, Johannesburg, South Africa, 2002.

tons of carbon emission per capita compared with the developing country average of 1.9 tons and the OECD average of 11 tons. Climate change may result in Mozambique losing over 25% of its potential agricultural production. Should this country be appealing for aid to the international community or should it be asking for reparations and justice, as it bears the brunt of climate change caused by others?

The world community of nations must fairly and equitably meet the challenge of addressing climate change mitigation and adaptation policies and measures. It must take stock of the differences between nations, their past and future emissions and must take into account prevailing socioeconomic conditions. The timely implementation of economically efficient and environmentally effective international agreements will be critical in the context of achieving societal goals of equity and sustainable development on a global basis.

IV. Political Responsibility –
Good Global Governance

Neoliberalism against Sustainable Development[1]?

Franz Josef Radermacher

$\Rightarrow$ Reducing the gap between North and South means more peace.
$\Rightarrow$ Technical Progress is the crucial instrument for bringing about acceptable humane conditions for more and more poeple on this globe.

Ever since the Rio Earth Summit ten years ago, if not before, the world has been faced by the challenge to consciously shape a sustainable development. This especially involves a great responsibility concerning the economic design. To do this while simultaneously trying to bring about a (global) social balance and also to adress the preservation of the ecological systems in itself a daunting and complex task.

This is a complex subject and the constellation has been increasingly dramatized by September 11, 2001, the Rio+10 Earth Summit in Johannesburg, which failed to a large degree, and now even more so after the course of events in Iraq.

A fair interaction between world cultures becomes a key issue here, if overcoming poverty with simultaneous attention to environmental protection issues and careful dealing with scarce resources should succeed. Even though technological and social innovations are an essential part of every solution. But these factors will not suffice alone.

---

1 This text is based on Franz-Josef Radermacher's book *Balance oder Zerstörung – Ökosoziale Marktwirtschaft als Schlüssel zu einer weltweiten nachhaltigen Entwicklung.* Ökosoziales Forum Europa, Wien/Österreich, August 2002
Sale in Germany: Herold Verlagsauslieferung GmbH, Oberhaching, Tel. 089 6138710, Fax 089 613871-20, email: herold-oberhaching@t-online.de
Sale in Austria: Ökosoziales Forum Europa, A-1010 Wien, Franz-Josefs-Kai 13 Tel. +43/1/533 07 97-0, Fax +43/1/533 07 97-90, email: info@oesfo.at
also can be ordered from:
Research Institute for Applied Knowledge Processing (FAW),
Prof. Dr. Dr. F. J. Radermacher
Helmholtzstr. 16
D-89081 Ulm

The Challenges of an adequate Global Regulatory Framework

In the new millennium, sustainability is a great challenge in terms of world policy. There is an international consensus that sustainability has to bring together two dimensions: on the one hand, protection of the environment in a global perspective, on the other, the development of poorer countries, especially with the aim of overcoming poverty and tackling other concerns of justice.

The key issue the world has been facing since the fall of the (Berlin) Wall is whether this aim is best reached by further deregulating the markets and then counting on the power of these markets. Or does this subject also require a framework of world economy, as it is typical for, i.e., European market economies? Namely, a framework which includes the ecosocial economic perspective in the sense of a liberal model combined with good governance, the model of the so-called Rhineland capitalism in Germany or the Ecosocial-Market-Policy in Austria.

It seems obvious that the development successes taking place in today's globalization processes are too costly. On the one hand, by a global massive destruction of the environment and, on the other, by an increasing social split in the North as well as in the South of this globe. This is not compatible with peace. This is not sustainable development. Here, the world faces a difficult situation which materialized, for instance, in the course of events on 9/11/2001 and also in the question of how to deal with this.

Exploitation Instead of Future Orientation

If one studies the challenges of sustainable development, one is confronted with the problem that the "sustainability" capital today, in other words, the social, cultural, and ecological resources upon which humanity depends, are seriously jeopardized in a globalized economy with an inadequate global framework.

We have been organizing international transport around the globe today almost free of charge with enormous negative consequences for the world climate. In the form of the green card, we have established mechanisms, which plunder the social capital of poorer countries. As a whole, this leads to instabilities that threaten future opportunities for life and development.

A large part of humankind, at the moment about three billion people, is extremely poor and has to live on less than two Euro per day. We realize that humankind is not in a position to guarantee elementary demands like water supply for all people despite scientific, economic and organizational potential.

A deeper cause for this seems to lie in the (essential) free trade logic of the WTO in connection with the mechanisms of action of the world financial systems. This is a regulatory framework dealing with social, cultural, and ecological questions on a subordinate level that brings them back to the level of the nation-states. In today's globalization, however, nation-states fight with each other over invested capital. In a certain sense they are in a prisoners' dilemma situation that forces everyone to reduce standards rather than to implement internationally coordinated standards.

This results in a comparatively uncoordinated, partly chaotic growth process with considerable societal distortions which, among other things, is characterized by the fact that it puts an enormous pressure on economically weaker cultures.

These cultures are put under considerable pressure as they are constantly lured by new opportunities, especially through the media. Due to their economic weakness, a large part of the population cannot avail of these. In its execution, the rich North at the expense of these cultures links this pressure with many material infringements.

This condition results in severe frustration and finally an enormous hatred that is understandable as it is an immense burden for coexistence.

Religions actually do a rule, as is sometimes implied, drive conflicts in the sense of a "battle of cultures." It is more likely that deeper lying concerns of justice that are not addressed correctly today, occasionally find their cultural demarcation line in religions to separate one side from the other. This is a function sometimes also taken over by skin color and by language.

Northern Ireland shows that such conflicts are basically not of a religious nature. In Germany, Catholics and Protestants live together very harmoniously. That is not the case in Northern Ireland. Why? Because deeper lying historical justice concerns are the real reason! Concerns of justice mainly affect the social area and the environmental situation on this globe, which is seriously deteriorating due to the working of the global economic system. The most important environmental issues for the future are water, soil, oceans, forests, climate, and the maintenance of genetic diversity.

Ecosocial Market Economy and Europe as an Example

The question is whether the globalization process has to proceed in such its current destructive manner. Or is there a better way? Yes, there is! There is an alternative. This is the European market model, the ecosocial market economy, the "balanced way."

The extension processes of the European Union as a small form of globalization are formed according to this logic. The next big step now is the extension of the EU to Central and South-Eastern Europe.

The crucial principle which the EU is counting on is a fair contract between developed and less developed partners. In this context, the less developed countries adopt the high standards of the EU (the so-called aquis communitaire) but also give up part of their competitive advantages. To put it differently, they save us from what we like to call dumping, which, seen from these countries' point of view, is their comparative advantage.

Such an adjusted course of action is only possible because the richer part of the EU is willing to support the development of these economically weaker countries with the help of co-financing. This more or less corresponds with the idea of a Marshall Plan the way it was conducted by the USA in Europe after World War II.

Comparatively modest financial means have to be used, to the order of 1 to 2 percent of the gross national product of the EU. Then it might be possible to considerably accelerate the catching up processes and to shape them in a particularly social and fair way. The obvious difference between this EU-type of approach and the Northern American NAFTA has to be pointed out here. In the latter, the border between the USA and Mexico has to be guarded by the military. Within the EU, borders will eventually disappear completely.

A Global Marshall Plan as Political Strategy

This idea of an ecosocial market economy should be expanded worldwide. This would mean that international agreements would link the alignment of standards, i.e., with regard to education, women's rights, water supply, and environmental protection to co-financing development of poorer countries by rich ones. Corresponding suggestions for a global Marshall Plan are at hand, especially from Europe's side. Co-financing is a central issue here. A fair taxation of international mobility, a world kerosene tax, a fair dealing with CO_2 emission rights, eventually a Tobin tax on financial transactions can be thought of in this context to raise the corresponding financial funds.

But today's problem is that the USA is blocking all the global processes of this kind even though the former Vice President Al Gore is one of the "fathers" of this idea and has also written a remarkable book on it.

The USA is not willing to substantially participate in co-financing. The richest country of this world only offers an absolutely inadequate development aid of around 0.12 percent (of the gross national product). The US military budget increase after 9/11/2001 alone is four times this volume. The 2003 military budget is 32 times the volume of the USA's development aid, thus 3.8 percent of the US gross national product.

It seems, US try to push the idea – if necessary flanked by the military and always with a military threat in the background – that only deregulated free markets are the best development program, even though it is quite obvious that global poverty cannot be overcome quickly this way and the environment cannot be sufficiently protected.

The recent substantial problems of the new economy with the world capital markets and the fraudulent redistributive processes regard to insiders that took place at the heart of economic and financial system have shown that a continuing deregulation is not the right instrument for the organization of classical economic processes, not to mention attaining sustainable global development.

The Role of Technological Progress: Factor 4 and Factor 10 Concepts

An ecosocial framework that suitably connects the possibilities of technological progress while observing standards in the environmental and social areas is more appropriate.

From the technological perspective, the most decisive approach is the so-called factor 4 or rather factor 10 concept that goes back to scientists like Ernst Ulrich von Weizsäcker and Friedrich Schmidt-Bleek of the Wuppertal Institute (described in the Indroduction of this report). This concept aims to multiply the world gross national product over the next fifty to one hundred years by tenfold. This would, however, only happen with a simultaneous increase of eco-efficiency so that this increased volume of goods and services could be produced without polluting the environment any further and without further consuming critical resources.

The idea is to produce more in the future with today's volume of resources and environmental damage, with the help of better technology.

Technological progress is the crucial instrument for bringing about acceptable humane conditions for more and more people on this globe.

Restricting Collective Human Behaviour as the Greatest Challenge: Overcoming the Rebound Effect

It is important to observe here, that an increase in eco-efficiency and dematerialization is not something new, as technological progress has always been able to accomplish this. Whether sustainable development can finally be achieved is another question. This requires other things besides technology.

Ethically speaking, social innovations, more precisely global contracts that set limits to collective human behaviour to keep it both within certain ecosocial as well as culturally acceptable limits are necessary. The imposition and implementation of such limits in today's world economic system is the true political challenge for a sustainable development.

If one looks at the climate issue and the challenge of a worldwide limitation of CO_2 emissions, the point is to limit the volume of collective emissions and thus achieve altogether fewer emissions than today. This, however, takes place at a time when China, India and Brazil are substantially catching up and, as a result, gradually producing more and more emissions because they are quite understandably emulating our lifestyle.

How is one supposed to deal with the scarcity and necessary limitations in this situation? Some very delicate discussions are taking place between North and South. The Kyoto Protocol, adopted by the Parties to the United Nations Framework Convention on Climate Change, in December 1997, is an important but not strong enough step toward reducing the threat of the global warming. Industrialized countries must reduce their GHG emissions by an average of about 5-7 % below 1990 levels by 2010. The reduction target for U.S. is 7% and the European Union is 8 % below 1990 levels.

The emission trading in the European Union starts now for practical experience from 2005 to 2008, when the emission certifications are given to the certain companies for free (grandfathering-system). The distribution will continue then in a bit different form till 2012. But it is not a global system and emission certifications for individuals are not foreseen.

This means it is most important to control the consequences of technological progress. Or, to put it differently, we need to prevent people from simultaneously using up more and more natural resources and producing more and more environmental damage as has always been the case historically in spite of technological progress and increasing efficiency.

In retrospect it can be said: "Die Geister, die ich rief, die werde ich nicht mehr los."[2] Technology has always opened up opportunities to relieve nature, but in the end, more and more people have increasingly polluted nature on an increasingly higher consumption level. This is called the rebound effect (see J. Neyrinck´s book "Der göttliche Ingenieur").

---

2 Citation from Johann Wolfgang Goethe: „The spirits that I called will not withdraw."

For attaining sustainable development, overcoming the rebound effect is the central global issue. This rebound effect can be found everywhere. Computers get ever smaller but the amount of electronic waste permanently increases. The so-called "paperless office" is the place with the highest paper use in the history of humankind. Despite telecommunication we travel more and more, not less. While we travel, we use telecommunication to organize the next trip.

This means technology is always only an opportunity. However, simultaneously incorporating necessary limitations into the world economic system through global contracts is required, to convert this opportunity into a solution.

The current free trade logic of the WTO is not in a position to do this. We have to develop the regulatory framework of the WTO further, in terms of content or rather, suitably link it with international agreements on environmental protection and employees, i.e., international agreements on the protection of children in regard to child labor.

Once again, this is bound to fail today because the poorest countries find it necessary not to have to obey such standards (even though they actually find them useful), because only this offers them chance on the world market.

If only the rich countries offer the poorer ones reasonable perspectives and co-financing, i.e., in accordance with the logic of the EU extension processes, it might be possible for them to consent to sign the necessary contracts together.

The Social Question as Key Issue: The Overcoming of Global Apartheid

Seen correctly, sustainable development today is mainly an issue of agreements that are required between North and South or between rich and poor. Environmental standards and regulations are important here for environmental protection. They need to be implemented globally and should be combined with the co-financing of development that would permit the poorer countries to still catch up economically in this process. Or to put it differently, this concerns a perspective for a global social balance in compliance with the observance of environmental concerns. Professor Klaus Töpfer from the UNEP is of the opinion that today the global social issue is one of the central issues for the attainment of a sustainable development.

In order to treat the social question one first has to determine how to measure the extent of social balance in countries. The EU logic takes the comparison of lowest income in relation to average income. According to EU logic nobody should have income lower than about half of the average (gross national product per capita) in the respective country. This corresponds to an equity of 50 percent.

This should be seen in contrast to extreme communism, which has an equity of 100 percent. Historically we know that a too high social balance does not really function. It is too discouraging and does not boost economic competitiveness. Instead, differentiations are necessary, including the possibility that certain major players can earn e.g. twenty times the average income, even if there should not be too many of them.

Correspondingly, it is inevitable that most people's income is below average. But how many and how much? If one looks at the successful states in this world, they all have an equity above 45 percent. Germany has about 57 percent, Northern Europe and Japan above 60 percent. The only successful country with an equity below 50 percent is the USA with about 47 percent. India and China are not much below that.

Empirically, all the successful, which means all the per capita rich countries of this world, have an equity factor of between 45 and 65 percent in regard to social balance. In terms of content, it can be explained why countries with an equity below 45 percent cannot be successful; why a country with too low social balance must be poor in regard to the gross national product per capita.

The underlying reason for this is that such countries cannot invest enough in the education and health of all citizens. Colonial or apartheid structures with a large number of service personnel on the lowest education and income level are the outcome and make a country very poor from a per capita perspective.

At this point, the neo-liberal argument collapses. If one takes socialist or communist societies as a basis, it may be true that the increase of inequity makes a country richer and is advantageous for (almost) all people. But, with a beginning of equity of about 65 percent, this statement is not true anymore. And, with equity below 45 percent, it is wrong.

Under such low equity conditions, it then becomes impossible to find enough qualified teachers or doctors who can educate the population and keep it healthy, etc. The result will be that in the end too many people will not have sufficient value added potential, at least not on an international level.

Those who do, cannot additionally also earn the money for their service personnel from a per capita perspective. Among the big states, the lowest equity factors can therefore be found today in countries which have a comparatively low gross national product per capita and still have conditions reminiscent of former colonial or apartheid regimes, as, i.e., Latin America (e.g., Brazil) or Africa (South Africa included) with equity factors of only about 27-30 percent.

Naturally this inequity also slows down growth in the long run. Truly high growth can only be permanently reached if equity is developed parallel to the extension of economic activities of a level of at least 45 percent. India and China therefore stand a better chance of becoming rich countries someday, as compared to Latin America or Africa.

The biggest problem however, is not the unfavourable conditions in most countries. Far worse is the state of inequity of the whole globe, if seen as an economic unit, which is perhaps the right way of looking at things.

The whole globe today is on an equity level below 12.5 percent. This is global apartheid: one which is significantly intensified compared to the former conditions in South Africa. This is an absolutely intolerable condition. It signals that inequities on this globe today primarily exist between and not within countries.

The world gross national product per capita today lies somewhere near 5,000 €. When the European definition of poverty is applied to the situation, no human being should have less financial means than 2,500 Euro per year, which means certainly not below 6 Euro per day. De facto three billion people are below 2 Euro, one billion among them even below 1 Euro per day. This condition is absolutely incompatible with peace and is also associated with hatred and opposition. The course of events on 9/11/2001 can be interpreted meaningfully within this context. It corresponds to the pattern of all previous revolutions in world history.

This is not supposed to mean that the poorest instigate revolutions or put up actual resistance. They are much too tired for this, too worn out, and too weak. Poverty and injustice, however, lead to other situations in which other persons at the heart of the system think they are justified to act as self-appointed representatives of the poor or act according to their interests. In this context, it has to be pointed out that 4,000 people died on 9/11.

From the USA's point of view, this justifies wars of aggression against comparatively weak states that are seen as a threat or rather are described that way even if they only have weapons and these could become a problem in the hands of terrorists. But it has to be called to mind that every day, 24,000 people on this globe starve to death. Since 9/11/2001, 24,000 people have starved to death every day. Even if most people and especially promoters of power in the North do not want to hear this, they are perhaps partly responsible for this starvation.

They have to endure whatever countries with a stronger economy and military impose on them. With mendacious arguments like "equal opportunities for all" (under a completely unequal starting point) they try to "sell" this as being just, which is intolerable and unjust – a double degradation.

But these players at the heart of the system should not think that people would ever accept this. Hatred builds up and looks for outlets. We should not be surprised with the results and the situation that emerge from it, especially with regard to terrorist and suicide attacks.

This is the greatest challenge for sustainable development. The continuing de-regulation of the markets alone does not bring the answer. Those wanting security in a globalized economy cannot put the responsibility for globalization's social consequences on the poor nation-states in the South of this globe. What is needed instead is a transition to global governance that is oriented in the way the EU organizes its extension processes.

We would then all be jointly responsible for social development worldwide and overcoming poverty and would be working to build up worldwide competitive infrastructures, to strengthen the woman's role, to establish educational and pension systems and so forth. This would lead to a condition where the global population does not grow anymore. The number of people would some day even shrink from the now foreseeable nine to over ten billion people for the year 2050, instead of continuously increasing, as has been the case up to now.

Peace between Cultures: Cornerstones of every Sustainable Development

The interaction between cultures and the cultural context as such are a significant part of the already mentioned (global) social issue. The differences within the framework of culture and culture reflects tradition. Tradition is, among other things, defined by what and how grandmothers and grandfathers pass on certain views on the world and on life to their grandchildren. Cultural imprints therefore go very deeply and are not easily changed. In addition, they can also be a potential source of psychological and emotional injuries, because of very deep sense of belonging, as well as tradition and expectations coming from childhood and youth.

Substantial cultural issues, among other things, concern the inter-generational interaction, as well as the relation between man and woman, and the public treatment of sexuality. These spheres of life have the highest humane significance and are partly tabooed in many cultures. Cultural memory lasts easily from fifty to two hundred years and more. At best, social changes of cultural patterns in these areas can succeed in a peaceful way over longer periods of time. Globalization does not allow such periods of adjustment any longer because of the tight economic link between countries and the global availability of information.

The washes like a tidal wave after a dam failure will come over unprepared societies. No one even asks if something might be wrong with Western culture any more. And this is so, though there is still no unequivocal and definite answer to the question of who is right on this globe.

Will backward, largely prohibitive cultures be the most sustainable in the long run or will it be the nearly unlimited free West where (almost) anything is possible? What is the effect of this unlimitedness in the West for social solidarity or sustainability?

If compatibility with peace is the goal, globalization processes should, at any rate, be shaped so that they support peace and balance between cultures and not aggravate conflicts. An integral issue is thus cultural balance and a dignified interaction with one another, no matter who is momentarily stronger or weaker in regard to economy, technology, or the military.

Money and power should not always decide which person, organization, or culture may assert itself in a conflict situation, even if this only means that the children of the "losers" are flooded with information and offers of a kind not permissible in the respective other cultural context. This also concerns subtle seductions or rather economic impacts which have incompatible effects upon the life patterns of the respective defeated culture.

To make the point even clearer, what happened with regard to the extensive extermination of the Native Americans and their culture in America or the enslavement and cultural rape of substantial parts of the African population during Colonialism owing to the superiority of Western culture in respect to economy, technology, and weaponry, should never happen that way again. Nor should it be done in a subtly concealed way under the guise of free economic processes that are formally directed towards equal opportunities, but do this on an absolutely asymmetrical basis which can never be fair because true equal opportunities do not exist a priori.

On the other hand, this also means that a reasonable global social balance, a (world) equity à la EU poverty definition would be a very important prerequisite for attaining a better balance between cultures than is the case today. All investments in a higher global social balance are therefore also investments in a higher cultural balance. This is true already for the reason that as a consequence of this higher balance other cultures are more able to economically defend themselves against the currently dominating Western model than before.

This observation has to be linked to another demand concerning social balance (one that exceeds a high equity), namely, that groups of people who are clearly separable according to categories like skin colour, religion, language, gender, etc., should be equally equipped with material goods. It is not very compatible with peace, if the world's poor are obviously amassed in some of these categories, with the result that belonging to one of these groups is already a poverty risk.

By way of comparison, it should be remembered here that a German federal state, i.e., Baden-Wuerttemberg has gone to a considerable amount of effort to provide balance between its two territorially and culturally separable parts, "Baden" and "Wuerttemberg," even though there are comparatively small material and cultural differences between them. And then one should think about the little effort being made for balance on this globe between a rich Western culture and an Islamic world that feels it is being treated unfairly and how much fuel does the West regularly add to the fire in a provoking and self-complacent manner.

Consequentially, this also means that purely individual-oriented human rights, as especially seen in Anglo-Saxon areas, are inadequate for a balance of cultures. Human rights should be seen instead in conjunction with human commitments, as described in a very nice book published by Helmut Schmidt. This way one would forge a bridge between the Western way of thinking and Asian philosophy which is much more oriented towards solidarity within groups.

The overemphasis that is placed these days on individual rights (i.e., the right of free movement) and requiring such rights from poorer countries, which are trying hard to develop, can be rated as eco- or resource-dictatorial aggression. This is a very subtle mechanism which helps richer countries prevent poorer countries from developing speedily by demanding the "impossible," namely, conditions which we also could not realize or pay for when we were on a similarly low level of development as these countries are today.

World Ethos and Fair Global Contract[3]

As described above, the main subject is a fair global contract which we have to agree on, if sustainability and compatibility with peace is to be reached. Such a contract has to be fair and find consent from all sides. If this is the goal, dialogues between cultures play a very important role. The contributions of the Parliament of the World's Religions and also efforts to formulate a world ethos, like Professor Hans Küng emphasizes it in the chapter "Sustainability and World Ethos", should serve as an example. In such discussions, common universal ethical principles are worked out, upon which all the major cultures and religions of this world can agree.

If one pursues this goal honestly, saving nature and the integrity of every individual as well as human dignity and equality prove to be the most important issues. One also has to prevent de facto double standards from being established, which happens quite often today.

A world ethical design is not an easy topic. Of course one will be confronted with extreme positions which cannot be tolerated under any circumstances, not even temporarily, like female genital mutilation or stoning of convicted persons

---

3 This is the special concern of Stiftung Weltvertrag (www.weltvertrag.org), where the author is chairman of the committee.

within the context of the Sharia in some Islamic countries. However, the West should take a look at itself as well. Not only the judicial system of the USA still apply the death penalty, it also applies it to children. Next to Somalia, the USA is the only country in this world that did not sign the world children convention for exactly that reason.

In the USA, we still also find a religious fundamentalism that is not only fighting actively against all UN measures taken with regard to the issue of reproductive rights, but also implementing "creationism" as an official alternative to biological evolution in school. The same question arises regarding to Israel's religiously justified takeover of the Palestinian land.

A discussion on ethical standards, for which a consensus has to be attempted on this globe, must also then name not just one but all the other fundamentalisms and also try to make changes there. This is true, at least, if the goal is to an order that can be accepted in the hearts of all people, including the major Arab states.

In any case, the good coexistence of Catholics and Protestants in Germany, for example, shows that a conflict like the one in Northern Ireland should essentially not be understood as being religiously motivated or impossible to overcome. This is not primarily a conflict between cultures (meaning a battle of cultures) or a conflict between two forms of Christianity. Rather, this concerns the overcoming certain unjust constellations defined in practical life sometimes by religion, sometimes by language, sometimes by skin color, as already described above. Islam per se is not a religion excluding modernization and secularization processes right from the start. The "hearings dimension" in Islam is a bridge that can be further developed for democracy.

Compared to other religions, the tolerance of Islamic states in the Middle Ages was exemplary. The incorporation of women in science in some Islamic countries took place much sooner than in the West. This means that there are obviously prospects for the further development of Islam and Islamic states towards a reasonable global contract. Work has to be done here as well for a world ethic. This is more onerous than a quick military strike and requires more intelligence, namely, empathy, which is the ability to abstract from one's own position and to

attempt to understand the other and also to learn from him or her. This means one should not be presumptuous and always know better, but be humble instead.

What is Necessary Now: The 10~>4:34 Formula for a Balanced Way

What is happening on the world order level? Of particular importance is to suitably link the WTO with other regimes and with other global regulatory systems.

Therefore, we ought to strive for an ecosocial consensus. If we tackle this in the right way, then we will actually have a reasonable ecosocial future. It is conceivable to transfer a growth factor of 10 over the next fifty to one hundred years towards a four-fold rise of wealth in the North of this globe in addition to the corresponding possible rise of growth of 34 times in the South of the globe.

The North would move from 80 percent of the "pie" to 32 percent of the tenfold volume of world economy. As a consequence of this development, the South could move from today's 20 percent of the "pie" towards 68 percent of the then tenfold larger global gross national product. This would be 34 times of the local gross national product. In growth rates, this corresponds to an average growth rate of 2.8 percent in the North, in the South to an average growth rate of about 8 percent over fifty years. This is better than today's ratios in India, worse than the ratio in China and overall not unrealistic. Countries catching up must primarily only copy and therefore can reach high growth rates. Countries at the top, rich countries, must invent innovations. In fact, fundamental considerations make it plausible that (honest) growth rates above 1 – 2 percent are hardly possible in developed rich countries.

The always surprisingly higher growth rates of the USA are, next to the indirect effects of purely speculative financial market bubbles, mainly a result of other accounting methods, ones in which technological progress is valued as growth way beyond market prices (so-called hedonic accounting). While this may be quite legitimate, if other countries do not do the same, comparisons are misleading. The discussed limitation in rich countries to a 1-2 percent growth refers to the fact that they do not have hedonic accounting, with the result that growth is assessed at market prices.

If we, however, reasonably combine worldwide both the high growth rates of catching up countries together with the lower ones (similarly high or even higher in absolute growth) of rich countries, we could find ourselves in a situation by the year 2050 where the people in the North are on a per capita average not twenty times as rich as the people in the South like today's "global apartheid." Instead, they would only be about twice as rich, which makes them on an average, four times as rich as today. This would be a level of balance à la European Union and certainly open up a perspective for a world democracy. This is not much different from the opportunity that arose from the EU's design processes offered to Europe by the European Convention.

In this respect, the ecosocial model opens a promising future perspective. It is an approach that takes human dignity and environmental protection equally seriously and parts with simple solution philosophies. In this view, a further increase of deregulation and social inequity will not solve the problems lying ahead of us. But hopefully, the activation of market forces under reasonable framework conditions of a social-cultural-ecological kind will. The author actually gives this promising program for sustainabilty only a 35 percent probability. What would then be the alternatives? This question will be dealt with further below after preliminary considerations on prosperity, growth, and social balance.

Prosperity, Growth, Social Balance: Some New Results

For some years now, the author has been working very closely on the connection of prosperity, growth, and social balance, mainly within the context of the EU project TERRA[4]. This has already been partially discussed. If prosperity is to be understood as a per capita high gross national product, a difference has first to be made between rich and poor countries. All rich countries in this world have a high social balance, more precisely an equity between 45 and 65 percent and they are all democracies. This situation concerns the possibility to secure excellent education, medical care, infrastructure access and equal opportunities for the overall population. This requires an appropriate number of well-paid specialists such as doctors, teachers etc. The result from these efforts is a social level of balance of at least 45 percent.

In rich countries, growth essentially only happens through innovation. Scientific research must be supported, innovations have to be carried out and implemented in markets. Democracies with a substantial support of scientific research and technology offer the best opportunities for this. Growth rates are reduced to a good 1-2 percent if hedonic accounting is not permitted. However, given the wealth of these countries, this is good enough.

The situation for countries catching up is quite different. These countries are comparatively poor. Partly, they do not have a high social balance. Since they are lagging so far behind, they can, in any case, reach high growth rates, by coping solutions and at the same time, involving increasingly more people in a formalized economy. Growth rates up to 10 percent are conceivable (leapfrogging), if not self-evident. A democracy is not really the most favorable structure for the organization of such catching up processes. Authoritarian systems like in Singapore, Taiwan or in China today can be of advantage, even though Japan has shown that at least under the specific Japanese democracy, high growth can be possible.

People can only become really rich, however, if a high equity exists, as it happened in Japan, Korea, and also Singapore, and as indicated in China and In-

---

4 TERRA (www.terra-2000.org)

dia. Democratic structures seem to be necessary at least when catching up processes are coming towards their end.

When taking this point of view, unlike China or even India, countries like Brazil and South Africa do not stand much of a chance of becoming permanently rich, unless the problem of social balance is solved at some point. In Brazil, among other things, another distribution of the land (land reform) must finally be carried out. To this very day, former colonial patterns of "above" and "below" continue to have an effect in these countries, just like in South Africa where, despite progress, old apartheid structures in the economic and educational areas still exist.

A certain moral dilemma lies in the fact that the rich of the corresponding countries are not always very interested in the increase of a per capita prosperity. Because of the much lower equity rate i.e., in a country with an equity of about 30 percent there are more people of a set absolute prosperity level than in a per capita country twice as rich with an equity rate of 60 percent. This means there are more rich people with more than ten times the average income than there are rich people in a rich country with five times the average income. There are more rich people in an absolute sense or in a relative one anyhow. Moreover, they profit additionally from very inexpensive personnel services, which actually cannot be financed to a great extent in rich countries with a high equity factor.

Under a corresponding inequity, the elite also have many ways of stabilizing their own position politically and intellectually by the use of financial means. Whereas the socially weaker side, which makes up the great majority of the population, is not even in a position to organize a corresponding intellectual counter-process even though under formally democratic conditions.

This can partly also be seen in the USA today. "The top of the pyramid" has e.g. succeeded in directing intellectual-political activity towards abolishing the estate tax. If it can manage to successfully repeal or substantially reduce the estate tax, substantial funds through think tanks and universities that are used by the "top" for political influencing would be invested in an extremely "value adding" way and would flow back to the rich investors with extremely high returns.

Ways to Disaster: Exploitation until Collapse or Resource-dictatorial Security Regimes

In the above, the ecosocial, sustainable global regulatory framework was given only a 35 percent probability of being successful in the sense of an ecosocial market economy and the question of alternatives was raised. Two alternatives are imminent. One of them is that we just go on like we have been and keep on abusing the ecological and social systems worldwide as much as we want. One day though, we will undermine the basis upon which our future and the future of our children depends. We will run into extreme scarcities, i.e., in the areas of water, food, and energy or in form of extremely high CO_2 emissions. There will be hell to pay as people try to secure scarce and too scarce resources and pollution rights for themselves in a race that no one will win in the long run.

This means that "ecologically we will be up against a brick wall" and the situation concerning environmental problems is uncertain. This is the panic scenario of all the greens and people who are environmentally aware on this globe. The author, however, does not believe this scenario to be very likely.

He believes that humankind, especially the rich world, will not be foolish enough to go on continuing in today's disastrous way, as this would destroy its own basis. From the author's point of view, this disastrous way has maybe a 10-15 percent probability. To put it even more clearly, in regard to ownership and property issues, the top of the pyramid is usually obsessive about property and takes hard and brutal actions against any development threatening what it feels are its legitimate property interests.

Historically, possible means, from lawyers, the police, even the military, has been used to protect property regardless of cost. The author therefore believes that things would not be any different, if it ever came to serious global resource conflicts or to conflicts resulting from environmental problems (i.e., in regard to the CO_2 problem). Although the position taken here is that we are not up against an ecological brick wall quite yet, this does not necessarily mean that we will come to a reasonable sustainable solution.

From the author's point of view, this does mean with about 85 percent probability that we will deal more sensibly with the problem of physical borders in the world economy in the long run than we do today. The author also believes that we will find solutions that will somehow integrate the last scarcities, in other words, physical necessities, into the world economic system.

The problem of how to avoid an ecological catastrophe is thus shifted to one that asks how burdens are to be distributed.

There are two possibilities left. One of them is the ecosocial way, a fair contract. This is what has been described above in detail and was given a 35 percent probability. But there is an alternative, at first unimaginable, but upon longer thought a quite obvious, seductive perspective, namely, an eco- or rather resource dictatorship in connection with a security regime. This third case is from the author's point of view the most probable (50 percent). Here, the rich North will some day deny growth to the poor South, like the rich prefer to do to the poor, just because a "business as usual" approach cannot be ecologically endured, if the poor do what the rich always have done. Here, the rich would especially have to hamper the development of the poor countries (i.e., relatively soon in China) or even destabilize them. Since all rich countries are democracies, we must ask ourselves if something like this is conceivable.

If one looks at what has been happening in politics in the recent past, especially the politics of the USA since 9/11/2001 and of the newer politics of Israel, elements of an eco- or resource-dictatorial security-oriented strategy can easily be supposed.

The USA's refusal to contribute fairly to the Kyoto Protocol is unmasking. This is even more true for the USA's almost fight against an International Criminal Court. Symptomatic is the USA's regular refusal to move within the framework of fair global contracts, as well as its persistent demand for special, strongly individual oriented human rights in poor countries which are unable to pay for all of this. The efforts at the OECD level point in a similar direction, e.g., to only give national guarantee for credits for investments in poorer countries, if products of the highest technological standards are bought. This way, if there is no co-financing, poorer countries are deprived of a great part of their competitiveness.

However, if the US was interested in helping human rights gain global acceptance, a global Marshall Plan would offer many possibilities to do this in a peaceful way. This would mean, that one would have to use one's own money for others, for instance in the form of a 1-2 percent co-financing of development. It seems money is preferably put into the own military and everything possible is done to secure one's own position.

If one compares the ratio of dead and wounded on both sides in the most recent Iraq War, the risks of the attacking side in this kind of war were very different.

One does not have to be surprised if the poor side, the weak side, does not find this fair and tries to defend themselves. How can "David" defend himself against "Goliath?" This leads to terror and even more terror and in the long run, the consequences are uncontrollable. Terror will be answered with even more national "counter-terror," followed by new terror, e.g., by suicide attacks. This kind of resistance is hard to fight and incidentally can cost us our civil rights and liberties in the defensive actions against terror. This process has already progressed a good deal in the USA. Suicide attacks require that people who think they are freedom fighters are willing to sacrifice their lives for a belief.

How wrongly organized must a world be, how much hatred must a world order provoke to cause such reactions? Is there nothing learned from this, i.e., about the violations done to others, maybe even unconsciously and unintentionally?

The rich North must seriously give thought to whether it wants to continue in its currently unrestrained way or whether the European model of balance in the form of a global ecosocial market economy is the better alternative. This costs 1 to 2 percent of the global gross national product for co-financing in form of a global Marshall Plan. As has already been mentioned above, such a plan has been suggested by former US Vice-President Al Gore. Basically, it is amazing how a chance for peace and sustainable development can open up so inexpensively with an intelligent instead of a dogmatic approach. It is even more surprising to see what kind of intellectual effort is made on the side of the biggest winners from the current deregulated structures of world economy to not pay this price and what kind of willingness exists to invest instead the corresponding means into more and more armament instead of humane developments around the globe.

Ecosocial Market Economy as the only Probably Realistic Opportunity

It is quite obvious that today's hope for a better future and a sustainable development is primarily put upon Europe and the developed Asian national economies. Together, we must win the USA over to another view on the subject. Therefore; we must be especially prepared to say that certain things are right and certain things are wrong, so that a lasting silence does not give the impression that we implicitly agree at some points. Seen this way, the principal objection to the Iraq War by the majority of the world was the just position. Within this context, the world civil society, the NGO's like Amnesty International, Medecins sans frontiers, BUND, Greenpeace, Stiftung Weltbevölkerung, Terres des Hommes etc, as well as the Rotarians, Lions, and other service movements have a big influence on forming of the global opinion, on facilitating understanding, and enlightenment in the truest sense of this philosophical position. It is important that these organizations speak out.

In the context, great hopes are pinned on the new information-technological opportunities of networking the world civil society which are used more and more efficiently. If one only succeeds in the struggle for a better world order to win over each year, one more person who stands up for a new, better solution and who at the same time wins over an additional person per year with the same belief and so on and so forth, then this snowball system will reach every human being in thirty-three years, as 2^{33} makes eight billion. When it comes to such an important issue, it should be possible to be convinced that one person per capita and year can be found.

This situation today puts a special political responsibility on Europe. Because of this, the introduction of the Euro was so important. Because of this, the extension of the EU is important. Because of this, the strengthening of the EU is important. In this difficult world this should also include building up the military strength of the EU in order to act independently with regard to these central issues of world order.

Is an ecosocial market economy a chance or a utopian dream? For reaching a peaceful sustainable future, an ecosocial market economy is probably the only chance we have and maybe the best innovation ever made in the political area, namely, the linking of reasonable balancing mechanisms and strict measures for environmental protection with the power of the markets and the potential of innovations. One can only hope that Europe, a continent with a difficult history and still incomplete identification and integration process, can take on the responsibility weighing heavily on this part of the world in this difficult phase of world politics.

Eco-social Market Policy as Political Concept

Josef Riegler

$\Rightarrow$ Use the mutual forces of economy, ecology and environment
$\Rightarrow$ The market – motor for sustainability

In January 2000, under the heading "Market Triumphs over State", the Austrian daily „Salzburger Nachrichten" made a remarkable comparison between the economic and political development at the beginnings of the 20th and 21st century: "At the start of the 20th century it seemed that market economy had finally won through. The prosperity gained through the Industrial Revolution was enjoyed to the full, but capitalism also showed its dark sides, miserable poverty contrasted frivolous wealth. At last, capitalism brought about its own collapse. At the end of the 20th century, world economy has gone back to the initial point of 1900, although on a higher level of prosperity. Neoliberalism is the enemy of the trade unions, globalisation progresses and demands sacrifices. It is up to the advocates of a liberal order whether they will repeat old mistakes or show that they have learned from history."

In fact, for the first time in the history of mankind, we are facing the question whether the entire human race will be able to survive in this new millennium or not – thus a focal point must be the quest for a survival strategy.

Two huge problem areas emerge:

1. Nature, Environment, Basics of Life – Sustainability

Since the mid 70s, humans have been using more resources than nature can reproduce. For the last 25 years, we have been increasingly living from equity and not from its interest. On the long run, such a development will not be feasible, will not be sustainable.

Whether in the Brundtland Report, in the Agenda 21 or in the EU's or Austria's Sustainability Strategy: "Sustainability" as a target is not being questioned.

However, we have to ask ourselves: How can we manage to shift from a civilisation of overexploitation to a civilisation of sustainability? Time is playing an increasingly big part here. Time is beginning to expire.

This is therefore the central idea of Ecosocial Market Economy's concept: We will only be able to manage the necessary transition to sustainability in the short time span available, if we do not pursue environmental protection against the interests of economy and consumers, but put market forces in the services of sustainability applying the "jiu-jitsu principle".

Ecosocial Market Economy as a Global Solution. The crucial idea is:

➢ to provide larger scope for economic and technological development by removing unnecessary barriers resulting from over-regulation, bureaucracy and tax burdens.

➢ to simultaneously strengthen the partnership principle and with fantasy further elaborated social fairness and, increasingly, enhance the symbiosis between public social policy and private initiatives.

➢ to integrate environment and nature into a price and cost framework and thus into business calculations for production, consumption, and transport.

In other words, we have to make use of the "basic instinct" innate to human beings to wish for themselves always the best and the least expensive – to make, by changing economic conditions, those things desirable for humans which serve sustainability. In concrete terms this means: renewable energy must be less expensive than limited and ecologically harmful fossil energy sources, renewable raw materials must moreover have more attractive prices than the limited stocks of mineral resources and oil; "throw-away products" must be higher priced and thus be of no interest to the consumers in comparison to those which have a longer life and are recyclable.

The concrete economic-policy tools within the concept of Ecosocial Market Economy sound rather simple:

1) to secure through a political framework and guidelines that environmental burdens as well as the consumption of valuable and limited resources are definitely integrated into cost calculations. This can be achieved by taxing water and air pollution just as by charging fees for the consumption of resources or by including costs of product disposal into the product price.

2) to carry out an „ecosocial" reconstruction of the tax system with the target to make the price of renewable energy and raw materials competitive and, on the other hand, to reduce the tax burden on labour and shift it to the consumption of resources.

3) to convert the use of tax income, in form of grants and subsidies, from the present system of subsidising overexploitation (nontaxability of fuel for aeroplanes and ships, subsidies for fossil and atomic energy, etc.) to subsidising innovation and business foundations for sustainable products and production processes.

4) to ensure that the sustainability principle can be considered in purchasing decisions by product declarations which are strict and transparent for the buyer.

However, implementing the above will require political courage on a national and international level.

These tools of Ecosocial Market Economy have to be introduced and implemented both within national measures and the European framework, within the WTO, global financial institutions etc.

The model of „Ecosocial Market Economy" is the only comprehensive economic and social concept presently known which seems apt to enable a qualitative and future-oriented development in the sense of sustainability within the framework of available economic-policy tools and international institutions.

2. Fairness and Peace

The constantly widening gap between

- poor and rich,
- haves and have-nots,
- power and helplessness,
- population explosion and stagnation

is very dramatic.

A global development which is exclusively driven by capitalism leads to straight out bizarre distortions:

- USD 1,500 Billion of daily foreign exchange trade face daily product exports amounting to USD 15 Billion;
- since 1980, the average holding period of shares has world-wide been sinking from 10 years to 10 months today;
- share of speculative foreign exchange operations: 97 %;
- 80 % of all cross-border financial investments are drawn back within a week;
- the income gap between the richest and the poorest fifth of humanity has been widening from 30:1 to 74:1 since 1960;
- according to an OECD estimate, the existence of tax havens causes USD 50 Billion losses in tax revenues.
- development aid from industrial countries to developing countries: USD 56 Billion;

– interest payments from developing countries to industrial countries: USD 135 Billion;

In a publication of the Austrian Federation of Trade Unions of spring 2001 („The Total Market") a striking statement was made:
„In market economy we are facing a total market and a social ideology which aims at economising all spheres of living. Neoliberalism, as propagated today, is nothing else than a reverse communism. Communism replaces the market with politics; neoliberalism replaces politics with the market. Both leads to a totality which contradicts European culture and a conception of mankind which recognises that the individual exists only thanks to the community and the community only thanks to the individuals."

In this respect and maybe surprisingly, Ludwig ERHARD, the father of Social Market Economy, formulated a matching finding:
„It is not the free market economy of the liberalistic freebooting of a past era, also not the „free interaction of forces" and like phrases, but socially committed market economy which allows the individual to establish itself, which puts the value of the personality on the top and awards performance with a merited revenue – this is market economy in a modern sense."

(From: "Committed to Society", Paul Bocklet a.o., Deutscher Instituts-Verlag, 1994)

Justice as a Prerequisite for Peace

The described misdevelopments can eventually be traced back to the fact that, due to the occurrence of simultaneous developments, politics was totally unprepared for "globalisation" and is presently standing aside rather helplessly.

Which were these developments?

➢ Politically, the collapse of the communist systems between 1989 and 1991 and the concurrent opening of huge new markets;
➢ Explosion of information technology and its widespread application, from PC over mobile phones up to the internet, which made financial transfers around the world possible within seconds;
➢ Explosion of capital markets without real global fiscal frameworks;
➢ Deregulation and free trade, boosted by the World Trade Organisation, without social and ecological frameworks;
➢ Dominance of the doctrine of „neoliberalism".

Blatant injustices and the feeling of powerlessness are an ideal breeding-ground for fanaticism, fundamentalism, terror and war.

DOHA – A Glimmer of Hope?

A decision of the EU Göteborg Council on the development of a EU Sustainability Strategy and the EU commitment for new rules within the WTO aiming at global sustainability are a first really weighty impulse to give the global development another direction:

A direction of social equilibrium, ecological sustainability and tolerance for different cultures, regions and mentalities.

The systems of thought of a "national economy" developed in the 19th and 20th century have only limited value for the challenges of the 21st century.

To cope with the future we need global rules for

➤ Economy and trade,
➤ Environment,
➤ Social Affairs,
➤ Financial and Capital Markets.

What can be definitely done?

a) We shall use the existing institutions as World Trade Organisation, UN Environment Programme, International Labour Organisation, World Bank, International Monetary Fond, OECD, etc., elaborate them further in the sense of Ecosocial Market Economy, and take care that not a single global agreement (WTO) can dictate to all others. It is essential that all agreements on economic, financial, environmental, and social affairs are equal in rank and value.

b) In a globalised economy and society, politics again must assume its role and task. It must determine the necessary rules of the game and set the policy framework for markets and human actions.

c) Most important for a sustainable development of a globalised society is the development of a world-wide ethical code which, as basis for human actions, shall be able to achieve consensus and serve as a necessary governance compass. The unbalanced cultivation of an "ethic of self-interest" has become a danger to humanity, because – if extremely applied – it also destroys one's own existence. The excessive orientation towards a pure materialism in the 19th and 20th century in both forms of capitalism and Marxism meant a dangerous dead end for humanity. We need an ethic of the "WE". We need a sustainable ethic, matching the holistic nature of Man as a being with body, soul and spirit and which – in a sustainable sense – makes "religion" as "reflection" of the "divine" possible again. Sustainable ethic requires the dimension of transcendence.

Einstein said: „We cannot solve problems by applying the same concept which produced these problems."

We need a new concept which roots in the wisdom of Creation and the experience of tried and tested ways to be able to shape the path of mankind in a good sense.

Creation's wisdom teaches us wholeness, diversity, networking, feedback. Which means that all what we do also concerns others, is connected to others.

From the experience of European development, we know about the importance of consensus and partnership, as the qualitative achievement developed in social market economy.

Both leads us to the model of Ecosocial Market Economy.

„Ecosocial Change"as a chance for a new quality in politics?

It must be a matter dear to the heart of all responsible citizens to concentrate the political debate within Austria on those moving forces which will be decisive for the life of an entire generation. We are also filled with the desire that the political debate in Austria is not egoistically bound in itself, but that Austria assumes a very active role for a long-term and just development on a European and global level in the sense of the Sustainability Strategy and Ecosocial Market Economy.

After all, we are talking about two huge challenges on a national, European, and global level:

These two challenges are:

"Sustainability" and "Ability to Live in Peace".

These two terms are closely related to the responsible treatment of nature as Creation and biosphere for many living creatures and many generations; with a fair distribution of chances in the field of economy, technology, education, health systems and sustainable social networks; respect among different cultures, religions, nationalities and different civilisations on the entire globe.

This respect for each other forbids an arrogant hegemony of one single political, economic and civilisatory power and requires priority for peaceful solutions to conflicts, as it has been practised by the European integration process for 50 years, with its principle of equal memberships, negotiation results and financial solidarity for compensation between poorer and richer regions.

Initiative for World Peace, Sustainability and Justice

In May 2003, on the initiative of the Ecosocial Forum Europe, representatives of different „non-governmental organisations" met in Frankfurt to bring a campaign for a world-wide Ecosocial Market Economy into life, whose aim is a sustainable design of globalisation.

In this process, among others, the following mutual finding was defined:

➢ The order of today's global economy is not sustainable:

Today's globalised world economy is nut suited to achieve supreme goals as sustainability and justice, as a global market lacks the necessary policy framework. Thus, a central concern is the necessary dovetailing of economic world order elements (WTO, International Monetary Fund, World Bank) with global social, cultural and ecological agreements and treaties.

➢ Ecosocial Market Economy as a model for designing globalisation:

A market which incorporates economic, social, cultural and ecological aspects in a balanced way is called Ecosocial Market Economy. Such a form of globalisation considers the requirements of a global ethos, as for example respecting the dignity of all human beings, establishing tolerance between cultures and protecting the environment. Ecosocial Market Economy has been the essential characteristic of European tradition in economy and society since the middle of the 20th century. It is also the core element of the EU's Sustainability Strategy and the EU's enlargement philosophy (cohesion, cofinancing between richer and poorer regions).

➢ Sustainable world order and global Marshall plan:

Derived from European experience, an ecosocial approach for designing globalisation is considered to be an apt basis of a development sustainable, peaceful, and future-oriented. The target is developing an efficient "Global Governance System" on the basis of a global Ecosocial Market Economy and a respectful behaviour within the "human family".

➢ Fair funding:

Following the EU's enlargement strategy, we need a fair system of global cofinancing to enable sustainable development in all parts of the globe and, on the other hand, to be able to demand respect of social, ecological, and human rights standards. As the US Marshall Plan for Europe once comprised about 1 % of the US Gross Domestic Product and the present common EU funding amounts to 1 % of the GDP, the financial volume of a global Marshall plan should also amount to between 1 and 2 % of the global GDP.

The financial means of development aid presently available do by far not suffice, nor can national budgets be expected to cover the necessary additional expenses. In the sense of a sustainability strategy and global fairness, it should be considered to raise new financial sources which make sense for a fiscal-policy system to supply at least part of these means for a global Marshall plan. "Logical" financial resources could be, among others:

Taxation of speculative global capital transfers, introduction of energy tax on fuel for aeroplanes and ships, trade with environmental certificates in accordance with the Kyoto Protocol; moreover a global agreement that throughout the globe a comparable taxation of capital revenues takes place to stop the nonsense of so-called tax havens.

New policy approaches for the future require confidence, power and courage. Such a policy also needs a spiritual climate and the necessary debate culture. A challenge for political parties and interest groups how to deal with each other, for the media, interest groups and lobbies and eventually for us all. In times of the internet, e-mails, talk-shows and readers' forums, everyone has a stage for shaping political ideas.

This is also a great chance for committed „non-governmental organisations" to contribute to sustainability and peace-keeping through networking and constructive work.

Good Global Governance: On the Necessity for Sustainable Awareness & Global Governance in an Interdependent World

Petra C. Gruber

⇒ Global Governance means regaining political control and creative power.
⇒ A global market will require an enormous extension of regulations, inspections and interventions.

Sustainability was once again to the fore at the World Summit on Sustainable Development (WSSD 2002) in Johannesburg. To this day, however, efforts to promote sustainable development on a global scale have had little success. Quite conversely, long-term difficulties have become discernible. Why is it not possible to make the overall significance of the concept of sustainable development tangible and to convey the message of new visions, of a new quality of a holistic life? Firstly, the unwieldiness of the term itself makes it difficult, since it implies some finger wagging. New endeavours are already underway using the new buzz phrase "responsible prosperity for all" – but it remains to be seen what type of prosperity is meant and who this prosperity is really for? It is said that one needs time for something to be constructive, but we cannot continue to rely on the associated cardinal virtue of hope alone. Additional measures must be taken to show everyone how vital sustainable development is to our very survival. Sustainable development cannot be prescribed. Sustainability can only be ensured gradually through a series of socio-political processes of concretisation and decision-making. A broad factual debate on sustainability will pave the way to establishing an appropriate social consensus. If all our senses are not addressed in this process, we will never dare the step from holistic awareness to the appropriate action. Innovative and creative approaches are required. The concept of sustainability challenges modern economic theories. Instead of banking on the growth imperative, it focuses on a dynamic equilibrium between the economy, social stability and the preservation of natural resources. The fourth, i.e., the political or institutional, pillar of sustainability, targets the sustainable development of "rules" for social coexistence. In a world that is characterised by ever more complex interrelations and interdependencies and vulnerabilities, it is no longer possible to resolve the global challenges, such as poverty, destruction of the environment, war, migration, unemployment, social conflicts, crime, international terrorism, and infectious diseases, on a national level alone. In order to control global change, there is a need for co-operation and coordination on all levels as well as interdisciplinary planning and holistic awareness.

Section 1 of this paper outlines the interdependencies between environment, peace and development in this era of globalisation[1]. Section II highlights the shifts in power, which benefit global players, as well as the basic regulatory conditions of Global Governance necessary for this. In addition to structural reforms and policy-making, the new holistic awareness explained in section III is a prerequisite for sustainable behavioural changes that build both on knowledge and values.

1. environment – peace – development as glocal challenges

Sustainable development as a normative concept conveys a vision of the world as it should be, and at the Earth Summit in Rio in 1992 it was declared an international objective[2]. On a macro level, the term sustainability was defined as the commitment to "equitably meet developmental and environmental needs of present and future generations" (Rio Declaration, Principle no. 3). The mission statement, which became known in 1987 through the Brundtland report, was based on the maxim of intergenerational action, in other words equitable action for all generations. Today, efforts extend far beyond focusing on the resources and sink issues and encompass the integrative and equitable treatment of the three dimensions ecology, economy and social affairs. Therefore, the fair intragenerational treatment, i.e., the distributive justice between and within the various countries, must also be embraced. All human inroads in social, economic and ecological systems should always be viewed under the aspects of responsibility and sustainability.

The long-term survival of the planet Earth and the world's population is jeopardised primarily by the inequality between people and their natural resources. Mankind's destructive impact on the planet has been felt early on. Historical examples can be found in antiquity already. Environmental crises at the time, however, were restricted to a local or regional impact. It is not until the beginning of the Industrial Revolution that we see the impacts of anthropogenic changes on a world scale: depletion of the ozone layer, climate change, decline of bio-

---

1 It is too narrow a defintion to equate globalisation with the internationalisation of economies. The removal of boundaries in daily life and behaviour can be felt in all dimensions of society, politics, economy, technology, and ecology. One therefore often refers to "globalisations" in the plural.

2 An historical overview of the development of the concept and development theories and policy has been examined in other contributions and would go beyond the scope of this paper.

diversity, deforestation, degradation of land and waters as well as the disposal of harmful waste.

The mechanistic worldview, which has been evolving since the 15th century, favoured the transition from an understanding of nature to a (supposed) domination of nature. The gap between man and nature grew steadily, and this isolation sowed the seeds for the "denaturation" of mankind. With the belief in progress, which had its heyday in 18th century Europe, mankind set out to "discover" the world. A virtually undisputed, optimistic and partially arrogant view of modern, industrial development prevailed until the sixties of the last century and still exists in some places to this very day. With the application of the economic paradigm to all areas of society, people were not only reduced to human capital or described in terms of their purchasing power or even cost factor, but nature too was exploited for economic gain as a source for raw materials or resources for the production process, or became a rubbish bin or sink for the disposal of harmful wastes. The neglect of nature's intrinsic value is causing the relationship between society and the environment to breakdown even further. This we find manifested as environmental damage and consequently as living conditions that are harmful to human health. The close link between environmental quality, health and also poverty is apparent. Health is not only essential to ensure a humane life, it is also a prerequisite to warrant the sustained social and economic development for all of society.

Most of the environmental catastrophes arising from our unsustainable life style do not respect national boundaries. Cyclones, floods and droughts render even more difficult the daily fight to survive in the poorest nations of the world, which are exposed to disproportionate environmental risks due to their more fragile eco-systems. Environmental problems are both the cause and consequence of (a broadly defined) poverty. The international community has – as one of the UN millennium goals – pledged to halve poverty by the year 2015. And poverty is not simply defined as a lack of material wealth. The definition of poverty as a non-fulfilment of basic needs for survival (food, water, clothing, housing and adequate sanitary facilities) does not go far enough – poverty also means cultural and social exclusion. Poverty means a lack of political participation and involvement by people in decision-making processes that impact them. And the repeated subjection to humiliation, exploitation and powerlessness associated with poverty are often exacerbated by a lack of self-esteem and poor self-confidence.

The last five decades of development aid have shown that it is not possible to import or export development[3]. Development is more than economic growth and technological progress. Yet as a result of the presidential address delivered by American President Harry S. Truman – when he assumed office in 1945 – four fifths of the world's population were thought to be "underdeveloped" from one day to the next. For the first time in history, entire nations were perceived as poor, or considered themselves to be poor, because they could not buy everything they needed to "live as human beings." People's behaviour and existence was overridden by the obsession of "having more." Traditional beliefs and the respect for Nature were "devalued" with one blow. Many felt and still feel that the purpose of development is to achieve the ideal image of a glorified (modern) industrial society as the highest level of social development. According to this linear evolutionary model, "non-Western" societies can and should be "civilised" by introducing Western socio-cultural, political and economic ways of life. This reductionist view of a global development ignores the many possibilities for culturally diverse ways of life and forms of expression.

Today's new transport and communications systems provide links between highly diverse cultures across national boundaries. Globalisation, however, is accompanied by a new trend towards relocalisation, a return to local particularities and regional cultural strengths, that is articulated in the compound word "glocalisation" (from global and local). Contrary to some fears, this is not giving rise to a globally homogenised way of life, a global "monoculture." The cultural exchanges taking place over the centuries have much rather evolved into a "global blend" and Western cultures are simply one aspect of this world society characterised by diversity and non-integration. This should not belie the risks and inequalities of new fundamentalisms and compartmentalisation trends however. The result seems not to be a global community, but rather "global apartheid." Only those people who have access to communication and transport are participating in the global world, for instance – this means exclusion for those who have no or only limited purchasing power. The poorest nations are becoming ever more marginalised – economically, socially and politically. Thus, one cannot speak of a successful world market integration. The three major competing global players

---

3 The term "develping countries" is not used due to the problems of the term "development." Instead they are referred to as nations of the southern hemisphere or South. Of course the North/South geographic distinction is problematic, just the generalisation of the "West." Industrial countries are now referred to as "modern industrial nations." The concept of underdeveloped has long been avoided on account of its associated with physical or mental inferiority and its humiliating undertones and the hierarchichal expression "Third World" has become obsolete ever since the end of the Cold War.

(North-America, Western Europe and Japan / South-East Asia) conduct three quarters of "World" trade, while Africa's share amounts to about two per cent. Thus, globalisation equates to marginalisation and fragmentation (or "fragmegration," i.e., simultaneous integration and fragmentation).

The General Secretary of the United Nations has confirmed that the majority of the world's population does not benefit from globalisation. The illusion of universally achievable economic wealth has faded for most people. Various attempts to achieve belated development have destroyed living conditions in many places, contributing to impoverishment and increased dependency, social and political exclusion in addition to the loss of cultural skills and considerably accelerating the destruction of the environment. The global propagation of the Western way of life, based on non-renewable fossil (and nuclear) energy resources, has last but not least revealed itself to be a "nightmare scenario" in view of the ecological limits. The first signs of a "chauvinism of wealth" can be made out in some areas – it calls on others to do some rethinking, while refusing to alter own lifestyles, but rather attempting to reserve these for own uses. The advocacy of environmental protection by rich countries is suffering from a massive credibility problem. This problem will not be solved unless we limit our ecological footprint and achieve a reversal to our excessive or erroneous development trends. Sustainability in the modern industrial nations implies above all awareness and a change in mentality and behaviour – "Living a good life instead of having much;" sufficiency is the dictate of the hour. It's not a question of asceticism but rather of frugality. There is more to prosperity or quality of life than material possessions.

According to Franz Nuscheler, the following key statement is crucial to help understand what co-operative development is about: "Development cannot be thrust upon you, development must come from within. One must underscore that development needs to occur in harmony with nature. Due to their reciprocal influence, it is no longer possible to separate man and nature. This awareness could be the impetus for the necessary move towards an integral and intact relationship between mankind and the environment based on a new responsibility ethic. External inputs such as funding, expertise and personnel cannot effect a self-determined and lasting development, they can merely promote such a development. Instead of charity and aid, which are rather vague and self-complacent, it is essential to finally allow other cultures to autonomously define and live the diversity of their own ways of life and to respect them in doing so. Each individual has his or her own creative and productive abilities and methods to resolve his or her own problems. Up to now, straightforward "entitlements to assistance" have placed people under tutelage and pushed them into (further) dependency, choking their own motivation. Consequently it is a matter of creating space, incentives and appropriate basic conditions so that a person's self-esteem, own initiative and sense of responsibility can evolve. However, it is important to examine closely whether the encouragement (or help) to self-help is not an easy excuse to refuse solidarity, co-operation and responsibility. Since the establishment of a self-supporting and sustainable system requires a long-term socio-political process, it would be fatal to suddenly drop people from a system upon which they were dependent. There must first be changes in the exogenous international political and economic factors that limit or disempower independent effort (such as financial speculation, agricultural subsidies, and other export restrictions, debt crises, etc.). Affected nations must also implement internal political reforms in their social and economic structures in order to increase their self-help potential. Thus African leaders, for instance, have taken a stance against poor government, mismanagement and corruption by founding NEPAD (The New Partnership for Africa's Development) in favour of accountability, transparency, democracy, and human rights.

The greatest threats to mankind lie in ecological, socio-economic and political grievances that lead to structural violence. Poverty, shortage of resources and their unjust distribution can provoke violent conflicts and present a huge challenge to peace policies. The causes of war are less likely to be the so-called ethnic reasons and more often about "valuable" resources, such as oil, water, timber, diamonds, or drugs. Apart from the humanitarian catastrophe, armed conflicts destroy life's resources, spread uncertainty and instability, and put the development of the affected areas back by years or even decades while forcing millions of people to become refugees. Hunger does not lead directly to terrorism, but coupled with hopelessness it creates a climate of violence. September 11, 2001 made manifest the link between global security and justice. Thus the impact is not only felt by neighbouring countries but has far-flung boomerang effects.

The structural changes necessary for global sustainability require co-operation on an international scale. There is, however, a large degree of reluctance to include the population of the southern hemisphere as partners with equal rights in global decision-making processes. In terms of economic and increasingly political and military power, the current relationship between North and South is one of force, superposed by a relationship of ecological dependency. Although it has long been apparent that the West needs further development in many respects and could learn from southern nations – for example in their relationship with nature or in terms of social cohesion – the transition from a one-way donor system to a global culture of learning is a slow one. Just, people-centred, socio-cultural and technologically adapted and ecologically viable, self-determined and interdisciplinary, sustainable development co-operation is not a humanitarian luxury. It is an investment in the future. Comprehensively understood sustainable development policies help protect and preserve our environment. They are the key to human security, for the peaceful coexistence of different world cultures. And they open up the possibility of a new holistic quality of life.

2. need global governance

New power relationships emerge in the course of globalisation processes. Under the dogmas of neo-liberal deregulation and privatisation, the economy has increasingly detached itself from society and reacts upon it now through (alleged) practical constraints. Nations that are defined by their population and their territory are subject to the control of transnational or cross-border players. Thus, transnational corporations (TNCs) play not only a key role in shaping the economy but also society in general and they can exert power without being accountable for what they do. In international (location) competition for investment decisions and jobs, nation-states and individual production sites are played off against each other (eco-dumping and social dumping). In doing so, transnational corporations "earn" much more from financial markets than from the production of goods. Nearly three percent of world trade is based on the actual trading of goods and services, the rest forms a bubble of speculative funds – the world acts as a huge company, a casino (hence the expression casino capitalism) without voter concerns or political accountability, but with frequently devastating and real consequences.

The repression of state participation either consciously or subconsciously imputes that markets basically provide a near-to perfect, self-regulating welfare mechanism for society. However markets are not in a position to correct themselves. The state warrants the functioning of the economy by establishing the infrastructural conditions necessary for investment and production as well as education. More market does not necessarily mean less state, but instead provokes an enormous extension of regulations, inspections and interventions. Privatisation, which is perceived as the panacea of economic policy, involves the replacement of state influence by private power (groups) – thus privatisation equates to a redistribution of power. Even if this is not immediately recognised as such, the result is a privatisation of profits and a socialisation of losses. The removal of state control mainly promotes the interests of those who dominate the market and who can thereby bar any competition – full competition is another fallacy: market competition is distorted, limited by monopolies or oligopolies, dominated by centres of power. The nature of this neo-liberal paradigm, which conceals the central ideal, debunks this as mere ideology. At its core, the "homo oeconomicus," i.e., the purportedly independent, objective, rational, unemotional, perfectly informed individual, who freely maximises own benefits irrespective of relationship ties or social connections, bears witness to an absurd, anti-social view of masculinity – and this one sided vision of humanity simply does not exist in the real world.

Today, entrepreneurs do recognise the challenges of the time, whether from a sense of social responsibility or in an attempt to prevent the system from collapsing. The economists Smith and Keynes referred to the necessary basic conditions and even Hayek writes about the need for embedding the market economy in its environment. We are faced with the great challenge to unite the market and humaneness and to bring them into line with nature. To do so will require the reorientation, reorganisation and strengthening of political creative powers. Josef Riegler's holistic model of eco-social market economy is an exemplary instrument for sustainability that manages to reconcile supposedly opposing economic, social and ecological interests.

The ever-growing gulf between poor and rich[4] within and between nations provokes social tensions, undermines political legitimacy and threatens peace. Global threats and challenges demand alternatives to nation-state political and democratic structures. This does not mean that the (nation) state has become obsolete. On the contrary, it has become indispensable as the only legitimate authority to protect public interests. Using the co-operative model of the transnational state it should be possible to re-inspire policies with new life (not only in terms of a nation state, but also in terms of a civil society). Shared sovereignty means gaining additional operational and problem solving capacities through co-operation.

Global Governance is necessary in order to co-operatively manage the core tasks for the future and the policy decisions of globalisation. It is not merely a matter of reinforcing international co-operation in international organisations but of a new political model that focuses on common welfare and also accommodates the emerging political areas as players along side nation-state politics and international regimes: transnational companies and players in the financial markets operating across borders, scientific bodies, the media as well as civil society and non-governmental organisations (NGOs), which fulfil an effective corrective function in world politics last but not least on account of their international publicity. Global Governance is therefore a complex process of consensus-finding and decision-making between governmental and non-governmental players from

---

4   The income of the poorest 10 per cent of humanity is equal to 1.6% of the richest ten per cent, who in turn earn the equivalent of the total earnings of 57% of the poorest people in the world.

a local to a global level in order to shape global change. The German translation for Global Governance – "Weltordnungspolitik," literally "world order policy" – may be misleading since it is not a matter of creating an international authority or world government. Governance without government is the new form of societal control as first envisaged by Immanuel Kant in 1795. A reformed UN system would be the institutional cornerstone of global governance and the reinforcement of a global rule of law would be the key building block.

Global Governance means regaining political control and creative power. The nations themselves remain the principal players, but take on the additional responsibility for coordination and reconciliation of interests of the various players with their divergent interests. Emphasis is placed on a sense of responsibility among decision-makers and their political will to implement multilateral agreements and national strategies through appropriate measures and to make available the required funding. Objections purporting that the common good is something rather alien to the prevalent thoughts and actions motivated by power and self-interests and that the chances of co-operation and a reconciliation of interests are rather slim must be countered by emphasising the need to co-operate on account of the urgency. The erroneous development trends and dangers, resulting around the world on account of the globalisation process, call for co-operative and coordinated action for reasons already explained – in our own interest or simply in order to survive.

It is still not clear how coherence between ecological, economic, social, and political systems can be "concocted" and effective, democratic decision-making structures can be ensured. Among other things, equal rights and possibilities to take action for all countries are imputed. Global challenges can only be met through a transformation of the current North-South relations – the maxim is "real partnerships instead of paternalism." In addition, international organisations need to be reorganised, the "institutional trinity" (The World Bank, International Monetary Fund (IMF) and the World Trade Organization (WTO) must be made more democratic, broadening their range of economic interest to bring about greater transparency and responsibility.

It is futile to build up a global order of finances and trade, a global social and environmental order if these are not integrated into a global order of peace. An international culture of co-operation, a "new spirit of the Global Village" should replace the old notions of opposing nations. It is imperative that the marginalisation of the United Nations, spearheaded by narrow-minded individual interests, and the unilateral hegemonic claims of the USA, be opposed by political alliances from "like-minded-countries" and from society in general. Security can only be achieved by working with and not against others.

3. and a new, holistic awareness

Basic conditions alone do not create sustainable world order. To achieve this, a change in awareness and the development or return of non-materialistic values is necessary. Sustainable living, for instance, is opposed by greed – this greed is self-propagating and insatiable by nature, dividing and polarising, perverting and devaluing, preventing love and contentment, greed alienates. What was once known as avarice has now become an economic virtue known as the maximisation of profits. The extensive application of the principle of competition, short term personal interests and so-called individual freedom are replacing basic human values and social achievements, and have detrimental effects in terms of the environment and quality of life for future generations.

Human rights are the basis for a comprehensive, positive concept of peace, as they call for key political, socio-cultural and economic "causes of peace." The following formulation of core elements for humane living conditions should also make clear that sustainable, peaceful development is a global challenge:

– Fulfilment of the basic needs for food, clean water, clothing and housing, including adequate sanitary facilities,
– Access to basic health services,
– Comprehensive education and equal opportunities in the access to information,
– Independence and freedom that imply the individual's responsibility for the common welfare of future generations as well as for the environment (as a value in itself),
– Cultural self-determination and the safeguarding of autonomous living spaces based on mutual respect,
– Personal responsibility, self-confidence and self-esteem,
– Socio-cultural and political participation, a constructive (political) conflict culture,
– Equal opportunities for men and women,
– Democracy and Good Governance, good government leadership,
– Non-violence and personal safety,
– Sustainable, i.e., socially aware and ecological, economy, and
– An intact ecosystem.

A healthy environment, freedom and justice for all, participation, personal responsibility and self-respect are "boundless values." A culture of peace is characterised by solidarity and hospitality. We tend to take our own value system as an absolute parameter. But the world order of others is on a par with our own.

The complexity of each way of life makes up the uniqueness and richness of this world. Dialogue, deeper intercultural understanding, empathy and mutual respect form the basis for a peaceful (international) reconciliation of interests and the secure coexistence of different cultures. It cannot be a question of equality or egalitarianism, but rather one of equal opportunities and fair distribution combined with a more holistic view of humanity. (See also Hans Küng's "Weltethos" [Global Ethic].)

There is little purpose in trying to assign blame or conducting reductionist analyses that seeks to play off economic causes against socio-political causes or individual mistakes. There is a far greater need for the development of holistic solutions. Specialisation in scientific fields has not really encouraged interdisciplinary discourse. In addition, the educational system provides knowledge that is removed from its overall context. Plato's understanding that knowledge and values are one seems to have been long forgotten. Originally, development meant the unfolding of human abilities and potential. A reorientation of educational policy to its original, comprehensive purpose will play a key role in this respect. An education that encourages independent thought and is geared towards the development of human potential could emancipate people from internalised constraints (e.g. consumerism) that impede the development of their consciousness and could grant them the scope to develop a critical relationship to whatever political and economic framework is defining their lives. Such a democracy-promoting educational policy is one of the prerequisites for personal development and freedom, for building self-esteem and for eco-social and political commitment – for genuine, pro-active individualism. This means not focusing solely on personal interests but also taking on responsibility, the emphatic inclusion and participation of people, based on the understanding that, in essence, man is a social being and can only find complete fulfilment and contribute to the common good within the framework of society.

We must break down accustomed, one-dimensional thought patterns and break out through the narrow boundaries of the traditional, mechanistic economic theory, in order to (re)establish the homo integralis. As part of the whole, mankind has a responsibility to nature; ethics are inherent to the bio-centric principle. If mankind achieves a deeper understanding of the overall interrelations of things, then mankind will voluntarily gear actions to benefit the recognised whole. Taking on the responsibility for the common good, including nature, is then commensurate with genuine freedom, far removed from the perverted individualism that is currently discernible whose own narrow remit prevents any advancement.

V. Economic and Socio-political Responsibility

Governance and the New Economic Order

Orio Giarini

$\Rightarrow$ The New Service Economy has to be thougt of as a global process involving the whole economy.
$\Rightarrow$ New economic theories must be adjusted due to the model of sustainability.

1. Introduction

Through Club of Rome achieved world reknown after the publication of its Report, entitled "Limits to Growth" in 1972, it suffered serious criticism. This was an important watershed: from the post-war period until 1972 the economies of most industrialized countries grew at approximately 6 per cent a year, while from 1973 to the present time the growth rate decline to about 1 to 3 per cent a year on average. The Club of Rome's doubts that "sustainable" growth could be sustained was regarded as a "breach of faith".

This article summarizes another point of view which states that 1972, a fundamental change took place in the way in which wealth was produced. The industrial revolution had been based essentially on investments in new machinery, tools and products. This had given way to the emergence of service-based economies. A series of reports were proposed through the Club of Rome to support this analysis based on over three decades of observing the manufacturing and the traditional service sectors.

The assumption is that the deterministic model, which still prevails in traditional macroeconomic analysis, has in fact given way to an indeterminate one. As a result, the key economic issue today is that of understanding and managing risk, the uncertainty and vulnerability of the economic system.[*]

---

[*] The following chapter is based on the book "Die Performance-Gesellschaft: Chancen und Risiken beim Übergang zur Service-Economy" by Orio Giarini and Walter R. Stahel, Verlag Metropolis.

2. The Legacy of the Industrial Revolution

2.1 Leaving Heaven for a World of Scarcity

One day, the Bible recounts, Adam and Eve were expelled from the Garden of Eden, and were forced to start a new life of labour and effort to survive. A world, though blessed with bounteousness, was scarce in directly available resources. Air for breathing was aplenty, but water for drinking and washing was not available in all places, nor was it always of the desired quality. Rivers and lakes then became favourite places for human settlements. Finding food could first be solved through hunting and gathering, but as the population density increased, this was no longer viable. Agriculture marked the onset of the first economic revolution. The descendants of Adam and Eve had learnt by then that most resources do not only exist per se, but also as on account of human knowledge and of an understanding of the human environment, as also because of the technologies humankind is able to develop.

Knowledge also enabled man to find new energy sources as substitutes for wood. Coal and petroleum have existed under the earth's surface for many millennia, but it took the development of chemistry and technology to harness these resources and to develop their derived applications. In fact, the reader of this report is almost certainly wearing one or more garments manufactured with fossil oil- based fibres.

The introduction of tomatoes and potatoes into Europe after the discovery of the Americas marked another stage forward in agriculture. It would therefore be quite wrong to imagine an ancient Roman enjoying a plate of spaghetti with tomato sauce! It was only a few centuries ago that geographical discoveries and technology development created a "resource" out of these commodities. A striking example of uncertainty in forecasting due to a new technology can be seen in the prediction Robert Malthus made in 1798. He forecast that resources would be insufficient to Europe's growing population. This "reasonable" prediction was defeated by the unexpected: the introduction of the common potato. It took in fact 150 years from the very first introduction of this crop into Europe to win over indifference and mistrust. However, by the beginning of the 19th century, the potato had gained wide acceptance and diffusion, especially in Northern climates where more traditional crops could not be grown. But even widespread acceptance and ease of circulation could not avert the great Irish famine early in the last century when the potato harvest failed.

126

2.2 Producing Tools and Goods to Increase the Wealth of Nations

Adam Smith was the first to lay the foundations of the theories for understanding and managing economic systems i.e. economics. Certainly economic analysis, and even economic theories, had existed long before Adam Smith. There are plenty of other economic observations in world literature. But it was Adam Smith, in 1776, who laid the foundations of economics as a specific discipline or science, distinct from more general social or historical analysis. So why was it Adam Smith? Indeed, during his lifetime, Adam Smith experienced the beginning of the Industrial Revolution, which witnessed the shift from an agriculture-based to an industrial-based economy. This transition is well illustrated by his opposition to the views of Francois Quesnay, Madame Pompadour's illustrious doctor and a physiocrate, who represented the French school famous for the saying: "laisser faire – laisser aller".

The dispute between Adam Smith and Francois Quesnay focused on the origin of the Wealth of Nations. Both had an explanation. For Quesnay, looking at the main source of wealth in France, it was obvious that the wealth of nations derived from a flourishing agricultural system. Adam Smith, however, was more concerned with the new developments in manufacturing he observed around him in Scotland. Since Adam Smith, the industrialization process has come to be regarded as the crucial weapon in the fight against scarcity. After all, Adam Smith was essentially a moralist as many other great economists, like Malthus and Marshall.

The Industrial Revolution is characterized by the appearance of distinct manufacturing processes, where there is a source of energy (the steam engine) which can propel a multiplicity of equipment (for instance weaving looms) and provide the mechanical impulse to produce the required movements (for instance pushing the shuttle containing the weft yams through the warp). It is at this point that the invention of the flying shuttle becomes feasible, increasing the rapidity and the precision of the shuttle's trajectory, which no longer needed to be pushed by the human arm.

The combination of a central, fixed steam engine with many flying-shuttle weaving looms requires the organization of a specific production space, a space which marked the birth of the modern manufacturing plant! Whereas in a primitive agricultural economy, weaving and any other type of activities could be performed in the home of the peasant in no specific time-frame, major advances in technology required, for reasons of efficiency, that labour moved to where the equipment was.

Furthermore, the concentration in production also meant that producing for one's own consumption began to diminish: specialization increased, a process accompanied by a need for trade. It was the specialization in manufacturing activities, and the growth of markets, which provided the empirical basis for Adam Smith's conclusion that the real wealth of nations can be built through the development of manufacturing i.e. industrialization.

The key to industrialization was the increase in productivity, i.e. the ability to use scarce resources so as to produce more goods with less resource. Specialized production technology and new and increasingly efficient tools (faster, consuming less labour and/or capital per unit) are key features of this process.

Industrial technology thus moved to centre-stage in the quest to increase wealth and people's welfare. Both culture and environment played a role in its development and its useful application.

It is important to note that the technological jump at the beginning of the Industrial Revolution was not a qualitative, but a quantitative, one. Technology has always existed in the form of tools since man first became active. One could equally apply the notion of technological performance to constructions in the animal kingdom (a bird's nest, for instance). Intrinsically, no major difference exists between the technology of prehistoric "engineers" who specialized in shaping stones in order to produce arrow heads or cutting tools, and the "engineers" of the first Industrial Revolution who developed tools, which, by contemporary standards, would be deemed extremely simple. In fact, most of the tools of the first Industrial Revolution are such that almost anyone, without having any particular university or scientific education, could probably reproduce with the tools available in most hardware stores. The "steam-engine" is in fact nothing more than a sophisticated system for controlling the increased pressure produced by a volume of water transformed by heat into steam in a given space. The common pressure-cooker, which many people now use in their kitchen, is based on the very same principle. The real problem is to produce the materials, recipients and related mechanisms, capable of resisting the pressure and of controlling its release. Similarly, the notion of the flying weaving-shuttle is very simple: the challenge was to produce a fixed hammer capable of hitting the shuttle with enough force to send it to the other side of the loom.

Only much later, towards the end of the 19th century, did the manufacture of tools and products start to depend on scientific knowledge, i.e. on the examination and understanding of problems and materials beyond the immediate perception of our senses. We know how to cut a piece of wood and we understand how boiling water transforms into a larger mass of steam. But we needed scientific research to discover that the same molecules that are, for instance, found in cotton fibres, can be reproduced in a similar – although by no means identical – way using oil as the raw material. Scientific research and the exploitation of technology based on science thus started to gain ground at the beginning of the twentieth century and they were professionally exploited only during and since World War Two.

Up to the middle of the 1920s there was no consistent investment in research laboratories in industry or elsewhere. Up to then, the cost of production could only be accounted for in terms of labour and capital costs. It is only since the 1930s that more and more money has been invested in research and development. Nowadays, research and investment, frequently with a 10 or 20 year span preceding actual production, can in some cases cost a company twenty-five percent or more of its total future sales income.

The Industrial Revolution witnessed tremendous progress, and it was a time of many discoveries and new technological "adventures". The main discontinuity has been the change-over from the sustained period of development of traditional technology that lasted throughout human history up to the end of the 19th century, to a new period in which the main, although not exclusive, impulse has come from the coupling of technological applications with the advance of scientific knowledge. This new process or marriage reached its peak after World War Two and has been responsible for twenty-five years of continuous high growth rates in most industrialized and industrializing countries. In terms of quantitative economic growth, this has been a unique phenomenon in the whole history of humankind.

The legacy of the Industrial Revolution as a whole must be regarded as a series of victories to increase the wealth of nations. In this quest the production of new tools and products in an increasingly cost-effective way.

3. Developing Capitalism

3.1. Money-driven economy

The second essential characteristic of the Industrial Revolution has been the monetisation of the economy. Money has, of course, always existed in some form, either as gold, silver or copper-coins, or as barter, as in the exchange of three goats for one horse. Indeed, until the beginning of the Industrial Revolution only a minor part of all economic activities had entered the monetary system.

In a pure agricultural society, the vast bulk of production and consumption does not enter the exchange system, where money has its origin. In fact, trade gives rise to money. Even if we take into account the glorious histories of the caravans which travelled Europe and the rest of the world or the numerous towns of Renaissance Europe which flourished as international market places for certain parts of the year, quantification will show that a very limited part of all the goods produced and consumed in those times were exchanged within a monetised system.

According to calculations prior to the 16th century, no more than 1% of the average life of a European was organized in a monetary system (the time spent in selling his time for money or using his time for trading). Today, the corresponding percentage would be well over 20%.

Also very revealing is the fact that in the old monarchies, the rulers often possessed little money since the latter was not an indicator of real power. The fact that banking activities could often be developed by marginal groups which did not belong to the upper classes, shows that up to the beginning of the Industrial Revolution money still played a secondary economic role.

In the past, money has always been linked to limited trading activities and, until the beginning of the Industrial Revolution, very little or no recognition was given to it as a means of stimulating production.

Adam Smith's importance must be recognized here, given the social weight of his moral convictions. In his book, *The Wealth of Nations*, he completely reversed the "moral" attitudes of the past centuries as depicted by Moliere. He clearly states that the God-loving person, one who avoids sin and endeavours to cultivate the most acceptable moral and social attitudes, is the person capable of saving.

Increased specialization depends on more trade which in turn requires more economic productivity. Greater availability of money makes it possible to save more and therefore to create capital for investing. This process has monetised the industrial world today on a vast scale.

As we have seen, the development of new moral and cultural attitudes parallels the emergence of new production processes and technologies. There can be no question that Adam Smith succeeded in making a virtue out of saving. Century and a half years later, for John Maynard Keynes indebtedness would in certain circumstances, – in a clearly deflationary environment – become a virtue rather than a vice.

Banks, which up to 1800 were mainly involved in trading, began to have savings and investment functions during the second half of the 19th century. Capitalists shared the joint ownership of new industrial ventures: thus the "corporation," or sharing of ownership, came into existence. Corporations grew and started to offer their shares outside a restricted circle of owners. Banks became the institutions for collecting savings from all sectors of the population. Subsequently, they began to function as intermediaries in channelling those savings towards productive activities.

It is important to distinguish the forms that monetisation took before and after the Industrial Revolution. Prior to it, monetisation was a relatively, marginal phenomenon. Its rapid development as an element essential to the functioning of the manufacturing process is however, typical of the Industrial Revolution. Parallel to this, a shift in power occurred as society moved from a pre-industrial to an industrial state.

When we speak of capitalism, we are merely alluding to the sociological and economic aspects of this phenomenon: the monetisation of the economy as an essential part of the Industrial Revolution. The Industrial Revolution, therefore, cannot but be capitalistic. The only important political question we need to resolve, then, is to what extent capitalism (the monetisation of economic activities) is compatible with, or even requires, a specific degree of political democracy. In any case, even a communist society is undergoing an Industrial Revolution they have to involve the concept of capitalism to some extent.

Nevertheless, today we can put forward some new questions: which type of productive activities (in a general sense) can be better stimulated through a monetised system and which through a monetarised one? These questions are dealt in the next chapter of Patrick Liedkte "The Limits of Monetisation".

4. The « Service » Economy

4.1 The Growth of Services in the Production of Wealth

The growth of service functions is the direct consequence of the development of production technology during the Industrial Revolution. Let us follow it step by step. Up to the beginning of the 20th century, new technologies and changes in production resulted mostly from improving practices in the workplace and through expired know-how. It was very rare that such changes or improvements were the consequence of research financed by a particular department or division, inside a company or in a research organization outside of the company. The professionalization of research only started during the 1920s as a reflection of the growing complexity of new technologies and the need to carefully plan their development and manage their achievements. This research service function, developed over the last 80 years, today involves millions of persons and substantial investments by both companies and state governments. Maintenance and warehousing of incoming raw materials and storage of products have always been part of even the most simple production processes. But the increasing specialization of production units, involving more and more complex and advanced technology, the growing need to protect the more sophisticated products against damage over growing transport distances, have, among other factors, contributed to the continuous increase in the cost of organizing such functions. At the same time, production costs have continued to decrease.

The distribution of products to more and more people in an increasing number of countries at great distances from the point of production requires the organization and operation of complex marketing functions without which the product simply cannot reach most of the consumers. The financial activities, as well as the insurance functions linked to the performance of production and distribution, become highly desirable and ultimately indispensable. When investment in one "machine" such as a nuclear power plant or an oil rig routinely needs billions of dollars, the need for all financial and insurance institutions becomes crucial.

As our society becomes more complex, so do the regulations governing human interaction, including product utilization and safety regulations.

At the beginning of the Industrial Revolution there was little need for a bakery or a textile mill to do any research in defining the qualities of its product and in targeting a market. Today, selling video-recorders, for example, inevitably requires detailed analyses of potential consumer profiles in terms of regional market differences, product pricing policy, age groups, etc. A variety of liberal professions, from doctors to lawyers, from market researchers and economists to consulting engineers, perform a large number of professional services, either within the production complex or as extern consultants.

Electronic engineers or physicists working in a laboratory are more educated than the technicians who operated the simple looms at the beginning of the Industrial Revolution, to say nothing of the great majority of labour functions performed before the Industrial Revolution which required a limited level of education. In the pre-industrial society, very few people in fact could, or needed to, read. In our services-oriented society, however, most people even require "computer literacy". Mass education expanded rapidly. Today, it constitutes a large sector in the industrialized countries.

In a modern economy, the health and national defence sectors are often larger than the realm of education.

To clearly understand and evaluate the modern service economy, one must keep in mind the production process itself. Technological development, which changed the production processes in order to enhance efficiency, led to a great increase in service functions during all phases of this process.

All the services mentioned are essential in planning, accompanying and supporting production up to sales, as well as products during their period of utilization. The "Service Economy", however, has brought to light another important service to be added to the list: waste management.

Concentration, specialization and increased levels of dangerous secondary effects are therefore the negative outcome of the use of more sophisticated and advanced science-based technology in various sectors. Parallel to the increase in industrial waste, the expansion of conspicuous consumption for an everincreasing number of people has also meant an enormous rise in the amount of waste produced by millions of consumers, in both quantitative and qualitative terms. A plastic bottle, unlike a piece of wood or paper, cannot always be burned since it may produce smoke of a corrosive or even poisonous nature. Devising a system for its efficient and appropriate disposal, therefore, requires additional investment.

Every product ends up as waste in the long-term! Most materials, including our own bodies, become waste at the end of their production and utilization cycle and some of that waste can be transformed into new raw materials. In some cases, this transformation process occurs naturally (as with organic waste), in others, only after a lapse of time involving human recycling. The recycling of waste is in most cases limited, either by "economic entropy" (when the cost of full recycling would be prohibitive) or by physical (absolute) entropy (when full recycling proves impossible for physical reasons).

The notion of value in a service economy is in essence linked to the value of any product (or service) in terms of its performance over time. It is this utilization value during the utilization period which is the key issue: For example, the effective performance (value) of an automobile as a mode of transport has to be assessed in terms of its period (and frequency) of utilization, and the effective benefit (value) of a drug has to be measured in terms of the level of health achieved. When in the industrial economy, the key question was: "What is a product's "monetised" value?" the service economy asks: "What is a product's "utilization" value; what function does it serve, how well and for how long?"

4.2 The Growing Industrialization of the Tertiary Sector

The development of a future service economy has to be thought of as a global process involving the whole economy.

In fact, service functions are integrated into all productive activities in the industrial as well as the agricultural sector. It is essential to note that new technology has forced the traditional service sector, also reffered to as the tertiary sector, to make radical changes to some of its functioning modes by introducing the processes which are very close to the capital-intensive processes in manufacturing. The distinction between the functions performed in a modern computerized office, and a control centre in a production factory, is rapidly disappearing. This fact has led some authors, when describing the characteristics of the contemporary economy, to speak about a "super-industrial" economy or a "Third Industrial Revolution" instead of a "Service Economy". These authors cite those sectors where technology is most advanced, and then point out that what is in fact happening is a process of industrialization of the traditional service sectors.

This is clearly an important phenomenon, but it overlooks the vast increase of service functions within the traditional productive sectors. The development of telecommunications, banking and financial services, insurance, maintenance and engineering, cannot be explained as merely new kinds of "production," a continuation of trends that previously occurred in textiles, as well as in the iron and steel and chemical industries. Selling textiles is a different business as compares to fulfilling a leasing contract over an extended period of time, during which the seller remains contractually committed to the consumer for the utilization of the product. The relevant issue here is to understand what the selling of an article of clothing in a service economy actually involves. In the case of renting chemicals, for example, means that you could return the used products for recycling or for disposal.

4.3 The Horizontal Integration of all Productive Activities: the End of the Theory or the Three Sectors of Economic Activity and the Limits to Engel's Law

Traditional economic theory still distinguishes between three sectors: the primary or agricultural, the secondary or industrial, and the tertiary which includes all services, sometime subdivided further to produce a fourth sector. This sectorialization is vertical and has produced theories of economic development according to which a historical transition from agricultural to industrial societies has occurred, and a transition towards a society where the service sector predominates could now be taking place. Such a theory focuses essentially on the pure industrialization process and on the predominantly agricultural societies, which were not seen as industrial. Thus, the tertiary sector is frequently not more than a "trash can" used to classify all those economic activities which simply cannot be called industrial.

In reality, for all three types of societies – agricultural, industrial and service-oriented ones – the relevant issue is the choice of priority in the production of wealth and creation of welfare measures. Even in an industrial society, agriculture could not disappear totally. Industry does not develop as a completely separate productive activity from agriculture, but influences the traditional way agricultural products are produced and distributed. In the same way, the Service Economy is not an outgrowth completely detached from the industrial productive structure, but permeates that structure, making it predominantly dependent on the performance of service functions within (as well as outside) the production process. The real phenomenon therefore is not the decline and growth of three vertically separate processes or sectors, but their progressive horizontal interpenetration and integration. In other words, the new Service Economy does not correspond to the economy of the tertiary sector in the traditional sense, but is characterised by the fact that service functions are today predominant in all types of economic activity.

Every fundamental change from one mode of production to another leads to a modification in the perception of needs and demand. The very definition of what constitutes basic needs also shifts.

In an agricultural society, agriculture was obviously perceived as addressing the problem of satisfying basic needs. After the start of industrialization, and in line with the history of economic theory, primary needs were defined in terms of the manufacturing sector's ability. Engel's law states that services are secondary in most cases because they only fulfil non-essential needs.

But, in reality, the true impetus towards the service economy was the service sector's vital role in making basic products and services. Services no longer merely constitute a tertiary sector, but are becoming the forefront of all economic activities, also in the industrial sector.

The insurance sector is a typical example.

4.4 From Product Value to System Value

Another key difference between the industrial and the service economy is that the first one attributes value essentially to the manufactured products which are exchanged, while value in the service economy is more closely related to the performance of a product itself and its real utilization (over a given period) integrated in a system. In the beginning, the value of products could be identified essentially with the costs involved in producing them. Now the notion of value in the Service Economy is shifting towards evaluation of costs in terms of the results obtained in utilization.

The first approach considers the value of a washing machine per se, the second evaluates the actual performance of the washing machine, taking into account not only its cost of production but also all other kinds of cost (learning time of those using the machine, maintenance and repair costs, etc.). The applicability of the two approaches is, in most cases, inherent in the technological complexity of the product: in the case of simple products and tools, the assessment of value can be limited to the tool or product per se. Nobody buying a hammer would think it necessary to take courses to learn how to use it. In the case of a computer, however, the cost of learning how to use it tends to exceed the purchase cost of the machine itself, especially when the former includes the cost of essential software.

Similarly, people buying goods such as dishes or even a bicycle would not consider signing a maintenance contract. With purchases of electronic typewriters, photocopying machines, or even television sets, however, maintenance contracts – even for individual consumers – are more and more common. In the Service Economy, it is not a tool that is being purchased, for people are buying functioning systems, not products. They buy performance.

System evaluation, i.e. the organization of tools and persons in a given environment to obtain desirable and economically valuable results, must also take account various degrees of complexity, as well as vulnerability, in system functioning.

Thus the notion of systems becomes essential in the Service Economy. Systems produce positive results or economic value when they function properly. The notion of system operation (or functioning) has to be based on real time and the dynamics of real life. Whenever real time is taken into consideration, the degree of uncertainty and of probability which conditions any human action becomes a central issue.

Any system working to obtain some future result by definition in a situation of uncertainty, even if different situations are characterized by different degrees of risk, uncertainty or even indetermination. But risk and uncertainty are not a matter of a choice; they are simply part of the human condition.

Rationality is therefore not so much a problem of avoiding risks and eliminating uncertainty, but of controlling risks and of reducing uncertainty and indetermination to acceptable levels in given situations.

Furthermore, the very systemic nature of modem economic systems and the increasing complexity of technological developments require a deeper and deeper economic understanding and control of the increasing vulnerability and complexity of these systems. The accident of the Siberian railway, June 5, 1988, when a leak from a LNG pipeline led to an explosion that destroyed two trains, killing all passengers, can serve as an example of systemic risks.

Unfortunately, the notion of vulnerability is generally misunderstood. To say that vulnerability increases through increase in the quality and performance of modern technology might seem paradoxical. In fact, the higher level of performance of most technological advances relies upon a reduction of the margins of error that a system can tolerate without breakdown. Accidents and management mistakes still happen even if less frequently, but their effects have now more costly systemic consequences. Opening the door of a car in motion does not necessarily lead to a catastrophe. In the case of an airplane, it will. This shows that the notions of system functioning and vulnerability control become a key economic function within which the contributions of, for example, economists and engineers must be integrated. In a similar way, problems of social security and savings for the individual have to take vulnerability management into account.

Thus, the notion of risk and the management of vulnerability and uncertainty become key components of the Service Economy.

4.5 The Notion of Risk in the Industrial Revolution and in the Service Economy – Moral Hazards and Incentives

Risk-taking was not studied in detail by the first great economists, because it was almost taken for granted by the cultural environment of the time. Schumpeter, however, made more explicit reference to the risk-taking entrepreneur. It was not until 1992 that the first comprehensive study of the subject was made by Frank Knight ("Risk, Uncertainty and Profit"). But even Knight tended to confine himself to a discussion of risk of the entrepreneurial type. The field of pure risk linked to the vulnerability of systems was still considered too secondary among the managerial objectives of firms.

4.6 Pure Risk

The activities of the service sector and of insurance in particular, have traditionally been regarded as secondary or marginal in the national economy, even though they have existed for centuries. Theories and even attitudes have not yet adjusted to the new facts in this field.

Some types of non-entrepreneurial risk are nevertheless now seen as more important due to changes in social philosophy. This applies to risks covered by social security and workers' protection in industrialized countries. Indeed, as early as the 1850s, the government of Prussia had organized the first compulsory insurance scheme for miners. But at the time of the great depression in 1929, this type of risk management was still in its infancy.

4.7 Risk of Entrepreneur and Pure Systemic Risk

It is important to insist on this issue: The connotation of risk in the Service Economy covers a much wider area than the notion of risk in the Industrial Revolution. With the latter, the main risk area involved was so-called entrepreneurial or commercial risk; in the Service Economy, this notion has to be extended to include so-called pure risk.

An entrepreneurial risk is one where the people involved in an activity can influence its purpose and manner by deciding to produce, to sell, or to finance, etc.

Pure risk is beyond the control of those involved in an activity. It depends on the vulnerabilities of their environment or of the system they work and it will materialize by accident, by chance. This notion of pure risk is exclusively related to the notion of the vulnerability of systems we discussed in the preceeding paragraphs and is a hallmark of the Service Economy.

One of the great differences between neo-classical economics and the new Service Economy is that not only is "entrepreneurial" risk taken into account (as in the case of Frank Knight), but that the notion of economically relevant risk is extended to include the notion of pure risk. The notion of risk, globally, therefore has two fundamentally different but complementary connotations.

Today, in any significant economic endeavour, equal strategic importance must be given to both types of risk (both being linked to the concept of system vulnerability).

Many people, in discussing risk management (meaning the management of pure risk) do not make a clear link with the global strategy of risk. Therefore, instead of showing how the two risks are correlated, they tend to confuse them.

The distinction between pure and entrepreneurial risk is also to be found in the notion of «moral hazard». This notion has long been familiar to insurers when they have had to face damages caused by those who have exposed themselves to risk for reasons of profit. Take for instance the case of somebody who burns down his own home in order to collect the insurance – the cause of more than 20% of fires.

4.8 The New Entrepreneur in the Service Economy

Managers and entrepreneurs in the service economy must be able to take a broad view of risk which embraces both forms (the entrepreneurial and the pure) of the phenomenon. Even the most advanced management schools today (Harvard included) are often lagging behind in this respect, for the reality of pure risk is beginning to impose enormous burdens on managers.

Risks have to be understood at all levels and controlled as to their level of manageability. Vulnerabilities can, and must, be diminished and checked. Only then can a strategic vision be developed and new challenges discovered.

Should their vision of the real world be partial or inadequate, both the entrepreneur and the public at large will be beset by the feeling of being overwhelmed by the risks and vulnerabilities of modem life. Yet that sense of powerlessness, of inadequacy, is rather the result of our cultural inability to identify, adjust to, and accept, the realities of our contemporary world. Thus, it is very much a question of attitude. This inability to adjust leads to pessimism and to fatalistic paralysis, like the sailor who, instead of using the winds to steer his boat, allows them to determine the direction in which his boat is pushed. It is crucial that we be able to identify these new winds blowing within the Service Economy, and that we recognize the challenges posed by the new risks, and by our increased concern for product quality and "use" value, for what they really are: opportunities for defining new directions, for stimulating renewed activity in our quest for real economic and social growth.

4.9 Tradability and Homogeneity of Services

Our difficulties in clearly stating the problem stem from the cultural or theoretical frame of reference used for analysis rather than from the problem itself. A particular point is the notion of the tradability and homogeneity of services. It is often said that an analysis of the Service Economy is almost impossible because services refer to such disparate fields such as hairdressing, telecommunications, or maintenance and health activities. But the same can be said of products; there is line homogeneity between a pullover, an airplane, orange juice and a watch. In fact, all "industrial products" are homogeneous only in so far as they are viewed from the standpoint of the production system, i.e. the manufacturing methods of production developed and improved by the Industrial Revolution. If one looks at services with an "industrial" mentality, one will inevitably discover that some of them can easily be assimilated to an industrial product while others cannot. However, the exercise is pointless since it tries to fit empirical evidence into an obsolete frame of reference.

The same problem arises with tradability. Many service functions are tested or considered in a way which assumes they can be fitted into the analytical framework developed for analyzing trade in industrial products.

Since the Service Economy is about producing results where the customer or user happens to be, it is clear that the notion of trade when applied to this context must alter radically. We can no longer distinguish between trade in services and the movement of production factors or investment as was the case in "industrial" economic theory. In many cases, trade in the Service Economy inevitably combines and confounds the two. For many companies, and especially those in traditional "service sectors", the equivalent of local or international trade in products is the organization of delivery systems where the customer is located.

4.10 Material and Immaterial Values in the Service Economy -The Value of Education

Numerous books and articles on the Service Economy (as well as on the "information" economy), have suggested that, in the present economic system, we are increasingly faced with so-called "immaterial" goods and values.

This notion of "immaterial" comes from the observation that during the classical Industrial Revolution, the production process mainly dealt with material (hardware) goods and tools. However, in our present service information society, goods are very often "immaterial" (software), as for instance an item of information or a computer programme (the support or transmission system remains "material").

The issue of the "immaterial" nature of services can probably be more usefully approached in the following way:

• there has always existed a combination of material and immaterial resources, in any type of economic activity. The fact is that during the Industrial Revolution priority was given (and in our view justifiably) to the material side of the problem: let us produce things first and later find a way to use them, for the world is dominated by scarcity;

• in the new Service Economy, in which material instruments and qualitative conditions of utilization are integrated (as, indeed, they have always been), the latter have become dominant simply because in the economic system, they now cost more (money and effort) than the mere production of tools. Therefore, there has been a shift of emphasis towards the notion of the function of tools (which is an "immaterial" concept, describing utilization) away from the earlier priority given to their material existence.

At the risk of repetition, it should be emphasised that in the Service Economy priority is given to functions, the primary concern being with result-producing systems. But it is equally obvious that these systems (even if they produce abstract artefacts like communications) are heavily dependent on material tools.

One should therefore be careful not to use the word "immaterial" to refer to a rather vague "idealistic" description of current economic development.

140

A "function" or a "system" is immaterial per se, just as a machine tool is "material" per se. The intelligence needed in both cases may develop in a number of different directions. More knowledge will come to the Service Economy as a continuation of the normal trend towards progress. Industrialization required a different level of investment in knowledge than traditional agriculture, but knowledge per se is nothing new: even the man who invented the bow and arrow was an "intellectual".

This takes us back to the problem of measuring the results as against the costs of production and of the absolute necessity of measuring value by some accepted indicators of personal and national wealth.

5. Value and Time in the Service Economy: The Notion of Utilisation

5.1 The Product Cycle: from Raw Materials to Recycled Materials

The "life" of any product can be divided into five distinct phases: design and conception; production, involving a transformation of natural resources; distribution (transport and packaging, marketing and publicity); the useful life over a variable period of time (the utilization period); and the disposal of the discarded good (recycling or waste disposal) This whole process is referred to as the Product-Life Factor.

– The fast replacement of goods has been a persistent trend in economic history, and has gained momentum in our fashion-based consumer society (the syndrome of bigger-better-faster new products), as economists have become preoccupied with production optimization, economy of scale and fast depreciation and replacement. The success of such industrial production has been measured in terms of flow at the Point-of-Sale (expressed, for example, in the GNP), while the notion of the use of a product over time, its utilization, has been largely neglected.

– However, it is precisely this utilization period which is the main variable in wealth creation! Who determines the length of the utilization period? A company can produce a plastic toy that breaks before it has ever been used and cannot be repaired, or a wooden toy that might last several generations, both with the same price tag and the same production cost and point-of-sale value. But how many of each will be sold year after year? Yet, the user has as much influence on the utilization period as the producer: identical goods such as automobiles, that are used in countries with different levels of development, will "last" an average of 5-10 years in "rich" countries, and up to 35 years in poor countries.

We have attempted to show that price is the yardstick, the reference criterion, around which we organize a measurement system capable of quantifying economic phenomena and results within the framework of the industrial process.

Price is given by exchange, and the money obtained from each transaction is then used to remunerate all those who have contributed to the production of that which is transacted, i.e. goods or services. Labours paid wages or salaries, and capital (representing an accumulation of labour in terms of tools made available for production, e.g. plant, machinery, systems, knowledge levels and managerial capacity) receives interest. Each contribution to the various steps of transforming raw materials into usable products or functions represents a "value added". Adam Smith built his notion of value on this idea of "value added" and considered it equivalent to the "exchange value". But the notion of value added has not just remained historical basis for economic theory. In recent decades it has become a reference for the fiscal system through the introduction of value added taxes.

It is essential to understand that the measurement of value added in economics refers to the measurement of a flow. Although reference is made to the selling price (which could give the impression that it is the measurement of a result), the reference to the cost of the production factors is conceptually linked to the measurement of what contributes to the production of wealth, and not to the measurement of wealth itself. This can best be explained as a bath-tub with two taps, as shown in Figure 1.

The bathtub contains a certain amount of water W representing a stock of wealth which we use for our needs and pleasure. This stock of water W is fed by two taps:

- tap M represents the flow of monetised production, which pours additional wealth into our stock W,
- tap NM symbolizes the flow of goods and services which also increase our wealth, but the production of which is monetarized. It refers for instance to free, unpaid human contributions.

When reading about economic indicators, many problems arise because of the lack of a distinction between what relates to our stock of wealth W (monetised or not), and what refers to the flows F (monetised or monetarized). The value added in economics is essentially a measurement of the monetised flow. It measures how much monetised production is passing through tap M to increase the stock of wealth W. The underlying assumption rooted in the Industrial Revolution is that any addition to the monetised flow represents an equivalent increase of the stock W.

Figure 1: The bath-tub of economic wealth.

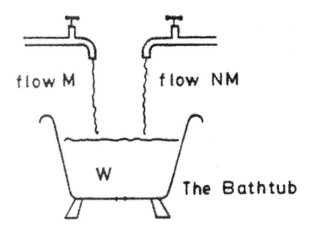

The measurement of growth as expressed in the Gross National Product is precisely and exclusively the measurement of such a monetised flow at the macro-economic, national level. It excludes the standard accounting practice used by all industrial companies.

5.3 The Bath-tub Systems: Measuring Results through Indicators

One of the major paradoxes in value accounting and in defining the development of wealth is that an increase in real wealth corresponds in some cases merely to an increase in the cost of pollution control (e.g. investment for waste-disposal and environmental purposes which is clearly a deducted value type of cost), while on the other hand, many real increases in value are under-rated. For instance, GNP growth figures published each year by governments indicate that the economy has grown by so many percent. However, a large part of this growth is in fact absorbed by factors which do not necessarily add to our wealth, while other factors that represent net increases in our well-being are not, or only inadequately, taken into account.

In the sense of the example of the bath-tub, it seems important to define a level for the wealth of nations in terms of stock, its increase, depletion, use, conservation and its diversification. Measurements of value added are important for the organization of an industrially productive system, which is an important sub-system of the economy as a whole. But it is only partially relevant to the business of measuring, targeting and organizing the wealth of nations.

Such measurements can be made using indicators which have been developed in many sectors and for many purposes over the past four decades. Yet, without the context of general economic theory, there can be non consensus on the definition of these indicators, nor can they be given the significance and status they require if they are to become efficient instruments for the general development of riches and of the real wealth of nations.

– Furthermore, the transition to an economic system and theory which go beyond the traditional notion of economic (added) value requires accepting a certain degree of uncertainty as far as measurements are concerned. This uncertainty stems from the fact that the very question of what wealth should be entails defining certain goals and expectations: the definition of a level of wealth is a function of time and history in evolution and, as such, a relative construct.

– However, where constraints are stronger, the need to survive is probably greater. Many potentially poorer people have in the past become more industrious and richer than those who inhabited a more blessed environment. This is as true for individuals as it is for nations. But it is a historical process and it can be reversed. Furthermore, not all advantages are necessarily species-specific, for where life is exuberant and easy, it is so not only for the human species, but possibly also for competing biological beings like viruses.

This whole domain is hard to define. Indicators of whatever kind, of the level of wealth, of health, of happiness, of knowledge and of the availability of material tools and means, are all concepts affected by uncertainty and change. The notion of value added appears to be much simpler and has the additional attraction of having been proposed and used as an instrument of universal management, as a standard that can be applied everywhere.

But the wisest way to proceed, in science as in other activities including economics, is it not always to start by using the simplest system?

– The problem is that the universal validity of the concept of value added resides essentially in its use as a measurement of an industrial production process. The establishment of a sound statistical basis for the measurement of the stock of wealth and its variation by means of an appropriate range of indicators which may differ from one part of the world to the next (but which do not preclude a minimum level of homogeneity for purposes of comparison), is not necessarily more complicated than the measurement of value added.

After all, there are already plenty of economic indicators in use which are periodically redefined, such as the consumer price indices that serve as a basis for the determination of the level of inflation in many countries. These indices contain within themselves a number of well-weighted elements.

– They are by definition not identical in all countries as they reflect the evolving structure of consumption. Why not define the real level of wealth or of riches in a similar way and allow the definition of wealth to vary much as the definition of the typical consumption pattern varies from one country to the next?

– In the mature Service Economy, this type of index might be politically more appealing, especially if it succeeds in closing the gap between measurements of GNP which do not reflect the reality of real wealth variations, and the perceptions of individuals, the "prosumers", who already have practical experience of what it means to become richer in contemporary economic conditions.

The Limits of Monetisation –
Changes in the Money-Driven Society[1]

Patrick M. Liedtke

⇒ A new yardstick for economic growth requires considering voluntary work as a contributing factor to the increase in wealth.
⇒ The concept of work has to be broadened.

1. Introduction

Unpaid work as a factor contributing to a growth in wealth has yet to be adequately reflected in the economic cycle. As a result, such volunteers often do not get due recognition, nor is the foundation laid to effectively shape the economic and socio-political development of our society. This contribution serves to reveal the underlying reasons, but also aims to show why things need not remain the same and how we can make improvements. These ideas are linked to future developments concerning our work. They are dedicated to sustainable prosperity which will only become a reality if and when we have understood and optimised all relevant dimensions of our actions.

The present text is based on the economics bestseller by Orio Giarini and Patrick Liedtke "Wie wir arbeiten werden" (ISBN 3-455-11234-X), published in German by Hoffmann & Campe, as well as in six other languages. The activities of the Applied Services Economic Centre (ASEC) in Geneva have been equally valuable. This non-profit organisation has set its sights on exploring further the interrelations within the new service economy and its future development.

2. Issues Involved in a Money-driven Economy

2.1. About the Wealth of Nations

As described in the previous chapter, economic science is a special discipline to be differentiated from other social sciences. As such, it was born on the very day that Adam Smith proposed an objective for economic activities, as well as a method of measuring the effort required to achieve this objective.

---

1 All rights beyond the publishing rights for this publication remain with the author

The common purpose was to develop the wealth of nations. This was to be reached by leveraging the growth potential of the industrial revolution, already well under way at the time. A high priority was to extend production capacities for material goods by increasing the contribution of labour and by using new tools. These were provided through the accumulation of capital and investments, based on abstract monetary units. This process could be measured because the contributions of the production factors (capital and labour) were compensated in the form of (essentially) freely transferable monetary units, which were independent of the actual production effort. To reflect this production process, a value was added, which in itself represented the measurement of a production flow. The generally held view was that the production of goods added value to the existing wealth, thus representing the most efficient method of producing wealth.

The way we conceptualise the economy today is generally expressed in the gross domestic product. Its growth rate from one period to another indicates the increase in economic growth and welfare. Evidently, this evaluation system has some serious drawbacks:

Goods that are available for free, such as fresh air, have no price. Hence, they have no value in purely economic terms. As a result, the destruction of freely available goods in the production process neither reduces the GDP nor the economic growth of wealth and welfare. However, it appears that it weakens factors which contribute to real prosperity and welfare. The real value of a good is not recognised until it has become scarce and hence, a price has been attached to it, irrespective of its original availability.

It seems paradoxical that a society which has limited access to drinking water and charges a price for it, appears richer than one in which drinking water is available cost-free. Indeed, it is the ultimate cynical consequence of our economic evaluation of growing wealth and welfare that a society which first pays for digging a hole, and then for refilling the hole, ends up richer than one that would have never allowed itself to become engaged in such an idiotic endeavour.

Especially at a time when for each process in the production of wealth, service functions are becoming more important than pure production functions, the traditional definition of value gradually loses meaning. Technical progress contributes to the development of increasingly complex systems and leads to a higher demand for services which become indispensable in order to make subsequent use of material tools or goods. It is from this angle that the increase in value in relation to the performance of a system must be assesed. This performance, in turn, may not be equated with the mere existence of a product and its (pure) production cost.

When looking at performance, e.g. human health, the fact that an individual is richer or in better physical condition than another can no longer be correlated to an increased number of medical products being bought and consumed. This also applies when a larger number of services become integrated into the consumption of a product.

From a historical point of view, a fundamental change occurred when for the first time, service functions became more important than pure production functions in terms of cost and resources used.

In order to integrate this idea into an economic model, refer to the terminology mentioned in the chapter of Professor Orio Giarini.

1. **Monetised** Activities:

They include gainful employment with all remunerated services or work-related activities. Hence, these activities are carried out in direct exchange for monetary compensation.

2. **Monetarised** (but not monetised) Activities:

This group includes all activities which are monetised and could be compensated with money but aren't as part of the transaction. Consequently, these activities have a specific market value. Based on this logic, "volunteering," "honorary post" work and "non-profit work" are considered monetarised but not monetised (as they are unpaid) activities.

3. **Non-monetarised** Activities:

This concept refers to all activities carried out independently of the production process itself or whose nature does not allow for their market-economy evaluation in the transaction, such as self-help, self-study or individual continuing training.

Another reason why unpaid work does not lend itself to economic analysis is that a systematic breakdown of this kind of work into clearly defined subgroups never occurred. The advantage of the above subdivision is that monetised work is systematically included in the official statistics (GNP, GDP, national income, etc.), with the exception of illegal undeclared work, whereas monetarised but not monetised work could be included. Only the contribution of non-monetarised activities, such as personal continuing training, to the wealth of a particular group of people cannot be measured using the available set of tools.

2.2 The Significance of Labour

When reflecting on the creation of wealth in our economic systems, one has to consider the fundamental functions of labour in our society. Although other production factors are required to create wealth, they are merely supplementing, albeit necessary, factors, to the core element which is labour. Any approach aimed at a sustainable system for creating prosperity needs to consider the functions of labour, which can be subdivided as follows:

1. The Production Function: Work enables individuals to accumulate the financial means necessary to earn a living.
2. The Allocation Function: Work ensures a re-allocation of resources available to society through re-distribution mechanisms.
3. The Solidarity Function: Work enhances social cohesion and fosters the organisation of communities through its social components.
4. The Function of Purpose: Work allows people to develop and express their value concepts – "We are what we do/produce."

Whereas the production function was the central issue in the past, the focus has been shifting increasingly towards the other four functions of labour. After a new wave of mass unemployment in the 70s and 80s, in particular the solidarity function and the function of purpose gained greater importance. Unemployment results in the loss of an opportunity to earn a living, a reduced feeling of self-worth, and the shrinking of human capital, which suffers from diminishing employability. It also leads to a decline in social skills which can, in turn, cause anti-social behaviour, as well as to a lack of social integration, leading to social division.

It appears that without the well-understood and efficiently organised monetary aspect of labour, unpaid, non-monetised activities have a serious drawback. Although they make an important contribution to the prosperity (economic well-being) of a society, they cannot claim future prosperity (economic well-being). This is chiefly attributable to the fact that no suitable instruments have been implemented and that society has not expressed its readiness to support this issue.

2.3 Monetised and Monetarised Activities

Up to the Industrial Revolution, most resources, which were primarily produced and consumed in the agricultural sector, were produced based on a self-reliant system: A non-monetary system. As we all know, the Industrial Revolution accelerated the specialisation process and consequently, the exchange process of goods and services using money as the medium. The money-based exchange process concerns all monetised economic activities. Keeping in mind the difference between monetised and non-monetised, a primarily agrarian society would be classified as predominantly monetised. On the other hand, monetised economic systems involve commercial transactions.

Although the history of money goes back a long time, the significance of money as an economic means is relatively new. As the industrial revolution progressed, money became the decisive instrument for the organisation of a new production system.

The successful development of productivity and industrial production has led to a paradoxical situation. At least as early as fifty years ago, Arthur Pigou, pioneer of the welfare economy, brought up one of the weaknesses of this hemisphere's economic system, without actually reaching further conclusions. He reflected on the fact that a bachelor who employs a housekeeper and then marries her, in doing so reduces the national income, as the previously paid work is no longer paid. Above all, unpaid work and the concept of non-monetised economic activities extend far beyond housekeeping and the non-consideration thereof which leaves a gap in national accounting.

It seems that in many areas, monetarised activities are relied on to save what appears to be in some cases the limits of efficiency of the monetised economic system.

In our service-based society, it appears that the connection between monetised and non-monetised (monetarised) activities is characterised by interdependence, and that a growing number of non-monetised activities actually constitute a form of productive work which is becoming increasingly important. This stems from the fact that these activities contribute to the prosperity of nations and in some cases, are essential elements for the monetised world to function in the first place.

The next move in the direction of a holistic system would be another significant step forward. Such a system would not only analyse and shape paid and unpaid work involved in an exchange process.

3. Changes in the Economy

3.1 Factors Influencing Future Developments in Europe

Which factors influencing future developments will, in our opinion, have decisive repercussions on the economic system? Once we have found satisfactory answers to these questions, we will be much closer to sustainable prosperity and steadily growing wealth.

In order to answer these questions more readily, the following considerations aim to sketch a plausible future scenario for the future development of our economic system, using five factors:

1. the structural transformation from an industrial to a service society;
2. the (political) harmonisation trends in Europe;
3. worldwide globalisation;
4. demographic development, and
5. technical progress. Broken down as follows:

• Whereas according to official statistics, around 50-60 per cent of the work force is employed in the service sector, currently, services already account for some 80 per cent of all work-related activities[2]—tendency increasing, as services have been spreading into the industrial sector. In fact, the IAB (Institut für Arbeitsmarkt- und Berufsforschung; En: Institute for Employment Research) projects a drop in the percentage of gainfully employed people working in the manufacturing sector from currently 16.9 per cent to only 12.7 per cent by the year 2010.

Enterprises formerly belonging to the second sector, such as IBM, General Electric, Ericsson, Nokia, Air Liquide or Schindler, now describe themselves as service-providers. Schindler in particular, a producer of lifts, seeing the future of the company in service-related activities, estimate that for the economy as a whole, the required percentage of people working in manufacturing will fall to a mere 8 per cent in the future.[3]

• The (political) harmonisation trends in Europe:

It is primarily the more recent rapid developments that have brought closer together the formerly very distinct legal, social, economic, financial and political systems in Europe closer together. This trend indicates that we can expect a stronger European integration in the future. In this context, the full implementation of the European Monetary Union will be sending particularly important signals. Today, many decisions are no longer taken on a national but either on a su-

---

2 See statistics of different EU member states (Cf. EU Commission). For a background analysis, see especially Gruhler, W. (1990): Dienstleistungsbestimmter Strukturwandel in deutschen Industrieunternehmen. The author describes in great detail how service functions have pervaded the activities of traditional companies in the industrial sector.

3 Currently, we are expecting an employment trend similar to that in the agricultural sector and are projecting a percentage of under 5% for the long term.

pranational level (see European Union or NAFTA), or, they are being delegated to and sketched out by large international organisations, such as the United Nations, the World Trade Organization, the World Health Organization, or the Organization for Economic Co-operation and Development.

• Globalisation trends:

In this context, globalisation refers less to the mere international exchange of goods and services but rather to the trend towards a stronger international dovetailing of value-added chains, based on efficiency and cost considerations. National economies are subject to external pressure and make national solutions more dependent on international circumstances.

• Demographic development:

In the case of the European Union, a significant shrinking and ageing process of the working population has been projected, a process which is discussed in detail in Chapter 9. In Germany, the population is expected to drop from some 82 million people currently to only 72 million by 2040,[4] a loss of around one eighth. In order to stabilise the population, a fertility rate of 2.1 children per female is required. This rate is not reached in any EU country; rather, the average rate amounts to 1.43. On the other hand, there is growing migration pressure from developing countries and countries in transition which have dynamic populations.

According to development forecasts for the labour market, the substitution rate of the labour force potential is projected at: -23 per cent for Germany, -28 per cent for Italy, -18 per cent for Spain, and approx. -5 per cent in France and the United Kingdom over a period of 15 years. By comparison, the figures for selected countries in transition are: Poland +15 per cent, Morocco +40 per cent, Kazakhstan +50 per cent, and Uzbekistan +125 per cent.

As regards the ageing of the population in Europe, it is projected that in particular the number of extremely elderly (80 years and above) will nearly double by the year 2040. The group of younger elderly (65 to 80 years) has been increasing as dynamically as the previous group; at the same time, the number of young employed people (16 to 39 years) has been declining significantly.

• Technical progress:

Primarily in the vast areas of biotechnology and medicine, as well as information and communication technology, we are currently undergoing rapid changes in terms of possibilities available to us.

---

4 Cf. the different population forecasts of the 8th Coordinated Population Projection by the Federal Statistical Office Germany, the OECD and various research institutes. The forecasts range between 67 million and 77 million people. Here, a median scenario was used. Additional numbers are generally based on OECD and ILO data as well as personal projections.

Over the last few years, the average life expectancy has been on the rise as a result of progress in medicine. In this context, particularly the elderly as a group have been able to benefit from a very significant increase in their remaining lifespans.

Consequences, Options and Activity Potential for the Future

In this section, the consequences of the aforementioned developments are described. These will be linked to the activity potential and options. We cannot determine precisely what repercussions the projected and described developments have or will have on the future economy and consequently, also on the potential for development in the area of monetised and non-monetised systems. Certain plausible consequences result from past experience. Ideally, these should be as consistent as possible with the goal of a sustainable creation of wealth. Whenever this is not the case, corrective measures need to be considered. The probable consequences, the way we deem the plausible, would be:

1. Interdependencies between national economies will intensify.

This will have an impact on the economic and social order. Historically, closer economic integration has also pressured those involved to adapt political and social institutions. Europeanisation and globalisation have an impact on work and vice versa. This means that on the one hand, the working population will have to face more external requirements and demands; on the other, however, it will also become easier to regionalise and integrate non-regional or non-national characteristics.

Unpaid activities, for which action is to be taken in the context of international requirements and interconnections, will not remain unaffected by this trend.

The migration pressure exerted by developing countries and countries in transition, which have dynamic populations, on more developed economic regions with stagnating populations will continue to increase.

At this point, I have just two more comments on this highly sensitive issue:

a) We are still not (at least in public debate) completely clear on what the dramatic consequences of this process on our societies will be. Both immigration countries, which have to deal with the influx of people, and emigration countries, which suffer from a "brain drain," have to prepare in a timely fashion.

b) Currently, we are not sufficiently prepared for the challenges that are developing. Germany, which has been a de-facto immigration country for a long

154

time, (still) does not have a tangible comprehensive migration law. In contrast, the U.S. has been pushing for "intellectual" and "entrepreneurial" immigration for some time.[5]

2. The significance of the production of material goods as a quantitative component both of the gross domestic product and of employment statistics will decline, whereas the relevance of services will increase.

Once this trend has run its course, only a very small proportion of the population will be employed directly in the physical manufacture of goods. In the previous chapter, the percentage of people employed in this sector in the future was projected to be 8%, and under 5% for the long term. This leads to particularly good prospects for growth of non-monetised activities. Linking paid and unpaid labour creates powerful synergies, especially in the service sector.

The "virtualisation" and de-materialisation process that is shaping work-related operations will continue and new types of work that have been made possible only by new technologies, such as telework, will gain in importance.

This trend also changes the qualification requirements for the working population, as the need for higher quality keeps rising.

In the 21st century, people are facing new and constantly changing work requirements, while work-related processes are becoming increasingly complex and abstract. Hence, they will push even more insistently for continuing training as an integral part of their work arrangements.

There are numerous signs indicating that the battle between companies for human capital, as well as that between companies and the work force over training demands, opportunities and cost, has just begun. Whether companies will be able to succeed in making the decisions concerning continuing training measures, or whether they will try to pass on more of the related costs to the employees, will depend on both the preferences of those concerned (companies and employees) and future framework conditions.

3. The growing independence of the individual "links" in the value-added chain thanks to technological progress will allow for decentralised work that will be independent of location. The reality of the working world is becoming a "virtual reality" as far as time and place are concerned.

---

5 Cf. the contribution of Laura D'Andrea Tyson, a former top economic adviser of U.S. President William Clinton, to Business Week (July 5, 1999). Under the title "Open the Gates Wide to High-Skill Immigrants", this is calling for exactly that.

Flexible working hours, which are beginning to show up in collective wage agreements and internal company agreements, are particularly prized by employees. New collective wage agreements are also more flexible than they were in the past with respect to the number of hours worked and how those hours will be distributed. Until a few years ago, the workload was always a central issue in collective wage negotiations on the number of working hours. Today, the times and places where the work will be performed become a second significant issue in such talks. This fact, however, makes non-monetised activities more difficult, since guaranteed freedoms tend to be curtailed rather than expanded, despite the potential reduction in overall working hours. This makes organising several activities at once more complicated.

On the one hand, there is partial retirement, which means that working hours are gradually reduced during the final years of employment.

On the other hand, the idea of partial retirement opens up new organisational possibilities. The Geneva Association has been supporting this idea for over ten years in the context of the well-established research project "The Four Pillars," especially as an additional element in complementing the existing retirement system and its financing.[6] It has been shown that older employees often prefer true partial retirement,[7] allowing them to work reduced hours (i.e., 15 or 20 hours per week instead of 35 to 40). They can feel integrated into the work process for a longer time, keep up their social contacts and tap into an additional income source to supplement the traditional pillars of retirement planning (government pension plan, corporate pension plan, private retirement plans and savings).

It appears that the trend is to further expand the idea of flexible working hours, which currently is still limited. Once this development has run its course, the time dimension will have disappeared completely from many professional fields.

Organising, and in the end, paying for the work time are auxiliary constructs that support our economic system, and are related to the experience of industrialisation, which dictates that productivity be measured as output per unit of time.

---

6 The Geneva Association (www.genevaassociation.org) publishes a semi-annual bulletin in the context of this research project.

7 It should be noted that there is a difference between the provisions described above and other arrangements which are also sometimes described as partial retirement but actually are more like early retirement (including the sudden transition from a full workload to none, as they imply that the workload of several years is compressed into a shorter period of time. One example is the arrangements made at the VW group, which were described as partial retirement but were actually set up like an early retirement programme.

The growing independence from location constraints has also fostered a development which, although welcomed by many gainfully employed, has also been criticised by others: the intrusion of work into their private life. For many people, especially women, the option of working from home makes it much easier – or even possible – to accept a job. Others feel uncomfortable about the lack of physical separation between work and their private life. As a result, many barriers between monetised and non-monetised work are reduced or even completely eliminated.

If paid work no longer stands in contrast to unpaid leisure time, and multiple commitments begin to overlap, then separation up along traditional lines no longer makes sense. The value of the work would then be re-defined, independently of the location where the work is done.

Such developments will also reinforce the imbalance in prospects for employment; regional and national borders will be overcome more easily.

In Germany, the gap in income distribution will increase; as opposed to most OECD countries, it has not grown over the last 20 years.[8] Those who are less qualified or whose qualifications do not meet the market's needs face a higher risk of unemployment, as well as lower compensation, whereas more competitive applicants have better prospects.[9]

5. In the future, the numerical dominance of younger workers will decline in relation to the number of older and self-employed workers.[10]

This will shift the balance of job design and planning and of the organisation of corporate activities in favour of the special needs of older workers. The rhythm of work will (have to) adjust to the needs of this group, which will end the dominance of younger workers on the labour market.

---

8 According to OECD data, the income gap in Germany from 1980 to 1993 between the second-highest 10% and the lowest 10% of all employed (based on the gross monthly salary of full-time male employees with a long-term job) has declined by some 8%. In contrast, the income gap in Japan, Great Britain and the U.S. has become even wider, with the gap in the latter two countries having increased by even more than 30%. According to the Centre on Budget and Policy Priorities, between the late 70s and the mid-90s, family income in the U.S. has seen a real decrease of 22% for the lowest fifth and a real increase of about 30% for the highest fifth.

9 Cf. OECD reports (1994): The OECD Jobs Study (especially concerning the connection between unemployment risk and level of education).

10 According to most population projections available for Germany, the percentage of people over age 60 compared to that of 20- to 59-year olds will increase from its current value of about 35% to 55% in 2020 and to 70% in 2030. At the same time, the average age in Germany will rise from 40 years today to 48 years.

In most countries where industrialisation occurred early on, the average number of hours worked has been declining in recent decades. According to OECD sources it has dropped—in Germany, for example—from 1,868 hours per year in 1973 to 1,724 hours in 1983, and down to 1,580 hours in 1998. Not only has this led to shorter times spent physically at the workplace; it also resulted in higher performance requirements on the job.

6. Companies' organisational structures will have to adapt to these developments and implement new forms of job design and planning that better reflect changing circumstances and the employees' wishes.

Even institutions that organise non-monetised work will have to come to grips with this development, as the potential work force also plays a decisive role in this context. The role of non-monetised activities in our society will continue to expand, because their importance for the creation of our wealth will increasingly be recognised. Experiences at the national and international levels, gained in recent years, have shown that attempts are being made to include social and charitable institutions, foundations, interest groups and non-governmental organisations (NGOs) into the process of shaping public opinion.

Risk and Sustainability

Walter R. Stahel

⇒ Insured damage is the visible part of the costs of environmental catastrophes
⇒ Cost for insurances increases – indirect internalisation of environmental costs

Risk-Management and Sustainability are two concepts that are closely intertwined. There are similarities, differences and overlaps. In moving from a 'protection focus' of command and control of the first two pillars of sustainability, to the 'innovation focus' of resource productivity, national policies to promote sustainability could be greatly enhanced by integrating the knowledge of risk management as balancing risks and chances. This could be crucial with regard to promoting new technologies in order to profit from the quantum leaps in competitiveness involved in many of the new technologies.

Up to September 11, 2001, the 'worst case' was ruled out in most risk studies. This meant that many States did not set the optimal priorities for their risk management, and that the biggest catastrophes were unconsciously accepted – without any measures of prevention. The events of 2001 led to the acceptance that risk assessment is a societal or moral problem, not a technological one. A better understanding of the mutual effects and influences of risk management and sustainability could be highly beneficial for society.

Studying the historic evolution of the concept of sustainability helps to understand its potential and limitations, and enables to identify the opportunities where the concept of risk management could improve sustainability.

The historical issues of a sustainable society

The concept of sustainability has historically grown on a number of techno-socio-economic issues:

1. nature conservation, or the eco-support system for life on the planet. This issue contains global system aspects (e.g. the global commons such as oceans and the atmosphere, biodiversity), as well as regional ones (e.g. drinking water, the carrying capacity of nature with regard to populations and their lifestyle, e.g. the footprint). This issue started with Jean-Jacques Rousseau 200 years ago.

2. health and safety, or non-toxicology. This is a qualitative issue, involving micrograms of materials. It is a matter mainly related to the health of people and

animals that are endangered increasingly from man's own activities, through the release of harmful substances into the environment, such as DDT, mercury, Thalidomide, to name but a few. This issue really took off after World War Two, but has been an issue of risk management for over hundred years.

3. reduced flows of resources, equivalent to a higher resource productivity. This is a quantitative issue, involving megatonnes of materials! It's objective is to prevent a possible re-acidification and a resulting climate change through, for instance, emissions of greenhouse gases, including CO_2 . Such a development would constitute a threat to man's life on Earth. This pillar is a child of the 1980s and 1990s.

The 'Quest for a Sustainable Society', however, must be much broader and include the longevity and sustainability of our non-technical socio-cultural structures:

4. social ecology, the fabric of societal structures: this pillar includes issues such as democracy, peace and human rights, employment and social integration, security and safety.

5. cultural ecology includes education and knowledge, ethics, religions and culture, as well as values of national heritage at the level of the individual, the corporation and the State.

Each of these issues is essential for the 'survival' of mankind within the natural eco-system – of which mankind is part. It is of no use to argue on priorities, or speculate on which of these issues can be lost first; society cannot take the risk of losing any single one of them.

Historic examples of sustainability, such as the Native Americans' rule of 'everything you do should have positive repercussions on the next seven generations', and the Prussian management rules for sustainable forestry, both at least 200 years old, were based on communities and their traditional values, i.e. a socio-cultural, not technological, ecology. They combined sufficiency with efficiency solutions that worked within their cultural circle – this was a bottom-up approach.

The principles of sustainability in modern times were formulated at international conferences at Stockholm in 1972, and reinforced at Rio de Janeiro 1992. At these conferences, the principles of intergenerational responsibility and a development compatible with economic, ecologic and social objectives were formulated with regard to the environment, but as a top-down approach.

The holistic vision of a sustainable society was also at the base of a business organisation that gave the term 'sustainability' practical meaning in the early

160

1970s: the Woodlands Conferences in Houston, Texas, and the related Mitchell Prize Competition. The industrial economy has been largely technology-focused, using monetarized values as its main yardstick. A sustainable society is result-focused and based on social and cultural values (non-monetary assets), integrated with economic values. Changing course towards a more sustainable society means to take into account social and cultural factors as peers to economic ones.

Identifying the problem – links between risk management and sustainability

The present economy is not sustainable with regard to its per capita material consumption in the industrialized countries. A dematerialization of the economy of industrialized countries can only be achieved by a change in course, from an industrial economy where success is measured in throughput and its exchange value, to a service economy where success is measured in wealth (stock) and its utilization value. Wealth management, new corporate and industrial design strategies, and different economic policies can lead to a higher sustainability as well as an increased international competitiveness due to a substantially higher resource productivity. The term 'service economy' refers not to the tertiary sector, but to an economy where the majority of value is created by services, and the majority of jobs are in service activities [Giarini and Stahel (1989/1993)].

The need for a change in course can also be explained in practical terms:

• The 'accumulation' principle with regard to the second issue: Paracelsus already recognised the fact that the difference between poison and medicine depends upon the dosage. There are a number of cases in environmental problems that suffer from accumulation: the CO_2-problem; heavy metals and non-volatile chemicals in the groundwater. Remedial action can only be taken against the dangers arising from accumulation by putting the precautionary principle into practice before a critical threshold is reached, and that is normally before scientific evidence of the danger can be given.

• The pressure of time with regard to loss prevention: A series of catastrophes, from the space shuttle 'Challenger' to Chernobyl and the channel ferry 'Herald of Free Enterprise', can be put down to the pressure of time. The simplest route to loss prevention is through a slowing down of the speed of the consumption of goods and a transition to a more suitable economy, without putting the competitiveness of the economy at stake.

• The pressure of money with regard to loss prevention: It takes a lot of courage in the present economy for a manager to stop for example the production of an oil-rig that daily produces 1.5 million dollars of income, or to spend slightly

161

more money to prevent a potential loss in the future – the catastrophe (in this case the explosion of 'Piper-Alpha') may after all never happen. A strategy towards loss prevention is put forward in the form of technological strategies for a transition to a fail-safe and self-curing technology, which is theoretically possible (e.g. built-in disengagement mechanisms).

• An important contribution can also come from the legislative body, e.g. through the introduction of an extended producer liability 'from cradle to cradle' like it was developed for electronic goods. The pressure of money (with potential catastrophic consequences) effects today the running of most infrastructures: in the USA, one street bridge collapses every day, mostly due to poor maintenance, and 180 tonnes of chemicals escape on average per day in 19 chemical incidents. A similar development is foreseeable in Switzerland or any other country, simply due to the fact that the maintenance of the growing volume and increasing age of infrastructures will impose increasingly higher costs, the approval of which is highly unlikely in the age of a tight budget and State budget cuts.

• The insufficient qualification of personnel in operation and maintenance: Switzerland recently, as one of the first industrialised countries, created a new profession, the maintenance worker. And yet, not least due to financial reasons, there are still insufficiently trained professionals running highly complex technological installations, which have been built and designed by highly qualified specialists. Catastrophes are thus unavoidable in many cases (e.g. a collision of two trains on the Gotthard line of the Swiss railway one Sunday in 1993, when only a trainee was on duty). The precautionary principle may be the solution promising the greatest success in these cases, as François Ewald quite rightly predicted (Ewald).

• Economies of scale versus dis-economies of risk: When looking at risk management of large risks it should also be taken into account that a coherence exists between the economies of scale in business (micro-economics) and the dis-economies of risk in macro-economics. In order to gain some business advantages, catastrophic risks are increasingly being accepted. As long as the last and in additional free insurer is the state (and this is the case wherever no unlimited liability insurance exists), there is little reason for industry to develop alternative technologies focused on waste-prevention and integrated loss-prevention over the whole product-life of goods, from cradle to cradle.

In addition, there are a number of the key issues of how industrial policy for sustainability could include risk management principles:

– Introduction of the factor 'time' into the legislation governing the economy;

• develop and use methods to measure sustainable competitiveness over long periods of time, e.g. GPI or ISEW instead of GNP [1].

• define and legislate the minimum quality of goods for sale in function of their service-life, by requiring a long-life warranty (e.g. in accordance with the 10 years in the EC safety directives), instead of the present exclusion of the use of used components in new goods.

• focus funding on R&D as well as education and training on prevention and precaution methods instead of process technologies: long-term behaviour of materials, components and goods (wear and tear versus fatigue), technical risk management, industrial design for system thinking, ways to popularise sustainability in terms of socio-cultural ecology.

– Increases in the self-responsibility of economic actors

• through a product responsibility 'from cradle to the next cradle',

• by replacing mandatory technical standards by mandatory free market safety-nets, thus introducing the insurability of risks as the main criterion for an economic choice between technological options,

• by eliminating subsidies and incentives for economies of scale which often hide diseconomies of risk.

• The use industrial policy as a locomotive for economic competitiveness: changing industrial policy ahead of the economic development in order to increase innovation towards sustainability.

• Sustainability as a holistic principle; the requirement of 'unity of matter' in legislation has become an obstacle to problem solving (e.g. levying taxes on resource consumption in order to finance old age and unemployment).

• Introduction of self-correcting taxation loops in the economy: tax resource consumption and waste instead of labour, thus rewarding all sufficiency and efficiency solutions (and eliminating the discrimination of 'voluntary work', neighbours help and 'non-productive work' at the same time).

Identifying potential solutions towards a sustainable society

The present industrial economy, which has developed over the last 200 years in today's industrialized countries, is based on the optimisation of the production process in order to reduce unit costs and thus overcome the scarcity of goods of all kinds, from food to shelter to durable goods, which was the norm 200 years

---

1 GPI General Progress Indicator (U.S.), ISEW Index of Sustainable Economic Welfare (European), GNP Gross National Product; for a details of GPI and ISEW please refer to [van Dieren (1995)]

ago. Emphasis is on more efficient process technologies, and a better quality of the goods at the point of sale.

There are indications that the industrial economy is no longer efficiently catering for our needs:

(a) the part of goods that go directly from production to disposal has reached 30% in some sectors, such as agricultural produce and electronics;

(b) the number of goods that are disposed of is comparable to the number of goods sold, for many product, indicating a substitution of wealth rather than an increase in wealth, in many affluent economies;

(c) technological progress is still focused on production processes, not on the utilization of goods and systems;

(d) for many goods, increases in efficiency through system break-down are comparable to increases in efficiency through product innovation (e.g. safety through traffic jams v. air bags).

And there are indications that the industrial economy itself is incompatible with the aims of a sustainable society:

● the factor 'time': sustainability is a long-term societal vision, concerned with the stewardship of natural resources (stock equals wealth) in order to safeguard the opportunities and choices of future generations. The industrial economy is a short-term optimisation of throughput in monetary terms. Changing course towards a more sustainable society means introducing the indeterministic factor 'time' into economic thinking, which again implies an indeterministic vision of economics.

● resource productivity: the linear structure of the industrial economy has as a consequence that its success, both micro and macro, is directly coupled with resource flows, of both matter and energy – it might be called a 'river economy'. But a generalization of the present per capita resource consumption of industrialized countries is not possible without a system collapse. In order to be sustainable, the economy must therefore operate at a much higher level of resource productivity, i.e. be able to produce the same utilization value out of a greatly reduced resource throughput – it might be called a 'lake economy'.

Changing course towards a more sustainable society means to de-couple economic success from resource throughput – one way to do this is to change to a service economy, in which the measure of results refers mainly to stocks and their utilization, instead of flows [Giarini and Stahel, 1992].

164

A service economy is fundamentally different from the industrial economy in that its main objective is to maintain or increase total wealth and welfare, i.e. the monetary and non-monetary assets of society, over long periods of time. Its focal point is the optimisation of utilization, i.e. of the performance and the results achieved with goods, rather than the goods themselves. The central notion of economic value in the service economy is the value of utilization over time, in contrast to the momentary value of exchange at the POS (point of sale) in the industrial economy (the added value system is only a subsystem of a larger economic concept). Similarly, quality in the service economy is defined as long-term optimisation of system functioning, not as a momentary quality at the POS.

The invisible hand of the free market prefers recycling to the economically more advantageous strategy of re-use. The reason for this 'wrong' choice lies in industry's familiarity with throughput optimisation, as well as in incomplete legislation. The economic optimisation of a loop economy only becomes an objective for manufacturers if legislated imposes closed liability loops, such as product take-back and product stewardship 'from cradle to the next cradle' by manufacturers, in addition to closed material loops. These liability loops are 'invisible' in a techno-economic vision and thus easy to overlook.

In case of the mandatory or voluntary take-back of goods, the new economic objective for industry becomes to maximize profits through the re-use of components and goods, instead of the old minimization of costs in recycling and disposal of goods. In some cases, manufacturers may need to develop strategies of 'retained ownership' (operational leasing, renting, selling results instead of goods) in order to guarantee the return of their goods after each cycle. Xerox 's marketing strategy of selling customer satisfaction in addition to its technological life-cycle design strategy is a brilliant example for asset management (for stock equals wealth).

Manufacturers of durable goods have a number of innovative technical and marketing strategies available in order to identify and optimise sustainable and economically viable solutions "from cradle to cradle", with the aim to provide customer satisfaction over longer periods of time. However, this also implies accepting a new definition of quality as a long-term optimisation of system functioning, as well as the increased product liabilities that come with it. It means facing new risks in a service economy!

Wealth without resource consumption? Facing new risks in a service economy

The objective of 'wealth without resource consumption' is obviously of little interest to the industrial (river) economy, as it will lead to "economic disaster" (as measured in resource throughput). There is therefore a considerable untapped potential of technical innovation and economic activity ahead for pro-active entrepreneurs that recognize and successfully develop this potential. The key to 'wealth without resource consumption' is the service economy: if customers pay an agreed amount per unit of service (and service equals customer satisfaction), service providers have an economic incentive to reduce resource flows, as this will increase their profits doubly: by reducing procurement costs for materials and energy, and by reducing waste elimination costs. Examples for this are the Xerox life-cycle design programme for photocopiers, the rethreading of tyres, the elevator business, Speno's rail grinding services, Du Pont's voluntary programme to take back and recycle nylon carpets, and the remanufacturing of goods in general.

Details of the main business and design strategies, which lead towards more sustainable solutions, are summarized in the following figure.

Figure 1: Demand and supply strategies for a higher resource productivity in the utilization of durable goods

Increased resource productivity through:	Type of strategies:	
	closing the material loops **technical** strategies	closing the liability loops commercial/marketing strategies
SUFFICIENCY SOLUTIONS (demand side)	near ZERO-OPTIONS ploughing at night loss prevention (vaccination)	ZERO-OPTIONS towels in hotels non-insurance (rear-end accidents in California)
SYSTEM SOLUTIONS reducing volume and speed of the resource flow	SYSTEM-SOLUTIONS Krauss-Maffei PTS plane transport system, skin solutions, accessibility.	SYSTEMIC SOLUTIONS lighthouses, selling results instead of goods, selling services instead of goods

MORE INTENSIVE UTILIZATION OF GOODS reducing the volume of the resource flow:	ECO-PRODUCTS dematerialized goods, multifunctional goods.	ECO-MARKETING shared utilization of goods, selling utilization instead of goods
LONGER UTILIZATION OF GOODS reducing the speed of the resource flow:	RE-MANUFACTURING long-life goods, service-life extension of goods and of components, new products from waste.	RE-MARKETING dis-curement services, away-grading of goods, marketing of fashion upgrades, goods in the market.

Source: adapted from: Giarini, Orio and Stahel, Walter R. (1989/1993) The Limits to Certainty, facing risks in the new Service Economy; Kluwer Academic Publishers, Dordrecht, Boston.

The main benefits of wealth without resource consumption for pro-active entrepreneurs are a higher long-term competitiveness through reduced costs, as well as higher product quality and customer loyalty, in addition to a 'greener' image; the main risk is the increased uncertainty due to the introduction of the factor 'time' into the economic calculation. The latter can, however, be substantially reduced by appropriate design strategies, such as modular system design for interoperability and compatibility between products families, component standardization for ease of re-use, remanufacture and recycling, loss and abuse prevention built into products.

Sufficiency is one strategy for higher sustainability and wealth without resource consumption. Witness a hotel: by offering its guests to 'save the environment' by re-using towels for several days, the hotel does indeed reduce the consumption of water, detergents and washing machines. But it also reduces its laundry costs and extends the useful life of towels and washing machines, thus increasing its profit margin. Zero-options, or sufficiency, are among the most ecologically efficient solutions – and they also offer the highest savings.

Systems solutions and the shared utilization of goods are also very effective efficiency strategies of higher resource consumption. A number of people sharing in the utilization of a pool of goods can draw the same utilization value through a more intensive utilization of a substantially reduced number of goods, thus achieving a higher resource productivity per unit of service. Examples for this are, besides public infrastructures such as lighthouses, roads, concert halls and railways, the Lufthansa car pool for flight crews, the 'Charter Way' concept by

Mercedes for trucks, and the textile leasing of e.g. uniforms, towels and hospital linen.

A shared utilization is possible in the (monetarized) economy through rental services and the sale of services instead of goods (laundry and dry cleaning), but also within communities (non-monetary) through lending and sharing. The former takes place within the legal framework of society, the latter's principles of sharing and caring is based on community values (trust and tolerance), which are part of socio-cultural ecology. Some of the issues involved in the sharing of immaterial and material goods are open to misinterpretation because they incorporate values of both society (law) and community (trust). Distrust leads to increased individual consumption, conflict or failure. A shared utilization of immaterial goods has two major advantages: a great number of people can profit from the goods simultaneously – in contrast to material goods – and: immaterial goods are by definition dematerialised. The technology shift from analogue to digital or virtual goods will further enhance shared utilization, even if the main reason for the change to virtual goods is competitiveness, not ecology. Product-life extension services of analogue (mechanical) goods lead to a regionalization of the economy, whereas digital and virtual goods enable producers to stay global, by providing solutions (for e.g. the technological upgrading of goods) through do-it-yourself activities. This gives producers direct access to the customer; it also eliminates distributors and distribution costs. The coming change to digital television, accompanied by long-life hardware combined with later technological upgrading through software, is an example of this trend – pushed by the novel German take-back legislation for electronic goods.

Wealth with less resource consumption is further possible by substituting maintenance-free long-life products, which deliver high-quality results for disposable products. Modern examples include music CDs (Compact Discs), and supercondensors (and in the near future rechargeable micro fuel cells) instead of batteries in electrical goods. CDs are also a point in case for the resulting shift in income from manufacturers to distributors (second-hand sales and rental shops), if the manufacturers themselves do not become service providers (e.g. selling music instead of CDs) – which would have demanded a structural change from global manufacturing to local rental services.

A longer utilization of goods through product-life extension services, as well as dematerialised product design, also increases resource productivity, but needs to be promoted as it goes against the logic of the linear economy. Doubling the useful life of goods reduces by half the amount of resource input and waste output, and in addition reduces the resource consumption in all related services (distribution, advertising, waste transport and disposal) by 50%. Furthermore, product-life extension services are often a substitution of manpower for energy, and of local workshops for (global) factories, thus enhancing social ecology. Economic success comes through an understanding of the logic inherent in a 'lake economy' based on services: to optimise utilization demands a proximity to the customer, and thus a regionalization of the economy. As the stock of goods in the market-place is the new focus of economic optimisation (the assets), these goods become the new 'mines' for resources. They cannot economically be centralized – an efficient service economy has to have a decentralized structure (service centres, re-manufacturing workshops and mini-mills). Service centres ideally are accessible 24 hours a day, such as the emergency department of a major hospital.

Benefits for the user cum consumer

'Service is the ultimate luxury', according to a publicity by the Marriott hotel group. The shift to a service economy (e.g. product rental instead of purchase) encounters few problems of acceptance on the demand side. The consumer turned user gains a high flexibility in the utilization of goods (something ownership can never give him), as well as guaranteed satisfaction at a guaranteed cost per unit of service. And there is no loss of status: the marketing of the industrial economy has wrongly created the idea that status symbol value is linked to ownership – in reality, it has always been linked to leasehold. The driver of a red Ferrari gets the same attention from bystanders if he has bought, rented or stolen the car. Ownership therefore only makes economic sense in cases where durable goods increase in value, normally through an increase in rarity, such as antique furniture, vintage cars, real estate.

And ownership only makes ecologic sense for individuals interested in asset management. In many countries, an increasing part of individuals live mentally in a multi-option society: they do not want to commit themselves medium or long-term, neither to goods nor people [Gross]. They want new toys all the time – and can afford them. Only a service economy can fulfil their needs without creating an avalanche of waste, by selling them results and services instead of goods, flexibility in utilization instead of bondage by property.

Most of these strategies of a higher resource productivity also offer the customer a reduction in costs. Sufficiency solutions based on a better (scientific) understanding of a problem reduce resource flows and costs: ploughing at night, for instance, reduces weeds and thus herbicide costs by 90%; remanufactured goods costs on average 40% less than equivalent new goods of the same quality; sharing goods also means sharing costs. But sufficiency and efficiency solutions often demand that the users cum customers develop a new relationship with goods and/or people – knowledge and community become substitutes for resource consumption.

Innovation and an industrial policy for sustainability as the keys to higher resource productivity

A fundamental change in actors and issues occurs when society evolves down the 'sustainability pillars' from 'health and safety' to 'resource productivity'. In the past, biologists and chemists have been the driving force through command and control regulations in order to conserve nature and limit toxicology, in the name of Nation-States. Now, engineers and industrial designers, marketeers and businesspeople will take the lead through innovation, in order to achieve an increase in resource productivity by a factor ten. 'Innovation by enterprises' and 'an industrial policy to promote sustainability' become the future key strategies not only towards a sustainable society, but also towards competitiveness!

This corresponds to a fundamental change in political thinking, from ecology v. economy (and State v. industry) towards ecology with economy (and State with enterprises). The new industrial policy can best promote sustainability by removing obstacles, which hinder, and by creating incentives which foster, innovation towards more sustainable solutions. The State still has to determine the need for safety barriers to protect people and the environment. However, the State should not provide this protection itself, nor carry the costs of accidents, but foster free market safety nets such as mandatory insurance (e.g. environmental impairment liability, product liability) [2]!

The State should define the aim of, but not the strategies to, a higher resource productivity. However, the State should make sure that economic actors who do innovate get rewarded and promoted, and that those caught cheating (or their safety-net) will pay up. By doing this, the State could become considerably leaner and more efficient. The principle of 'insurability of risks' would automatically introduce the precautionary principle into the economic mechanisms to choose between possible technologies, present and future.

---

2 'free-market safety-nets' are economic actors which can give a guarantee that financial losses resulting from innovators' mistakes do not have to be borne by the State: pools such as P&I clubs in shipping, insurance companies (including captive insurers, reinsurers), 'Berufsgenossenschaften' in Germany. These losses can arise from environmental or product liability, workers safety and compensation, etc.

Benchmarks for sustainability

Experience has shown that the resource efficiency of existing solutions can be improved by up to a factor four (e.g. water use through drip irrigation in Israel, use of herbicides by railways). For an increase in resource productivity well beyond a factor four, however, innovative strategies for new solutions are needed, attacking problems on a systems instead of a product level, or departing from a new understanding of the underlying need, or using new technology [Stahel 1995].

The discussion on resource productivity has shown the need to measure old and new solutions not against alternatives but against optimums. Otherwise, most of the praise might go to the worst actors of the past, whereas existing ecologically optimal solutions will go unnoticed, or even be forced out of the market by aggressive new-green marketing. Many traditional solutions have indeed reached a high degree of sustainability: an Austrian cabinet maker that uses timber from the local forests to produce furniture and toys for the regional market, repairs products when broken and heats his workshop in winter with waste timber, can hardly improve the sustainability of his activity even by a factor 4. The same goes for a local brewery in Wales, which buys its raw material from local farmers and sells barrels of beer to the local pubs. The fact that these firms can hardly improve their ecological efficiency does not mean they are working in an unsustainable way – quite the contrary! But it shows the necessity to establish benchmarks in order to define priorities and objectives in the quest for a higher resource productivity.

Benchmarks are easy in domains where people can be used as a yardstick, such as mobility. 'Sustainable mobility' can be defined as any method of mobility, which enables a person to move faster and with less energy input than by walking, such as the bicycle for horizontal and the elevator for vertical mobility. The chances, however, that improvements to the motorcar will ever reach the human yardstick are low. In other cases, such as the coffee brewed by different coffee machines, the optimum benchmark is the quality of the result achieved. During the last coffee tasting competition organised by the EU consumer associations, one of the best result was achieved by the Bialetti espresso machine, an early eco-product designed and first produced in 1930 which is still on sale – together with hundreds of other coffee machines, all more expensive, more recent and more material intensive. Benchmarks can therefore also be used to indicate areas of eco-mature solutions.

Similarities, differences and overlaps of the two concepts

Similarities

The common denominator of the two concepts, risk and sustainability, is that they are cultural constructs.

Risk is a construct on safety and security,
Sustainability is a construct on quality of life.

This cultural base means that both concepts enjoy a broad diversity of ideas, which in turn indicates that they can hardly be globalized without a loss of this richness of diversity.

Both concepts are founded on prevention strategies: loss prevention in the case of risk management, and the precautionary principle in the case of politics of sustainability.

This mental foundation on the principle of prevention often leads to a clash with reality. For the concepts of risk and sustainability both contradict modern economics, in terms of practice and theory:

Road accidents, to take but one example, lead to increases of GNP, as ill health and damaged automobiles lead to an additional economic activity that would not have existed without the accidents! But few will claim that the quality of life increases with the number of road accidents in any society. Accident prevention through risk management thus increases sustainability (ecology improves if fewer cars are smashed up through resource savings and prevent waste, economic savings are achieved through less expenses for health care and new cars, social hardship as a result of accidents is avoided).

The fact that successful prevention enables a society to maintain existing social wealth (good health) and economic wealth (well functioning goods and technical systems) is not taken account of in today's economics, which measures throughput (GNP) but not riches (total wealth available), as detailed in the chapter by Orio Giarini.

Differences

Risk management and sustainability differ considerably in their objectives and liabilities.

As far as objectives are concerned, risk management reflects its origin in manufacturing and is primarily concerned with wealth optimisation, whereas sustainability reflects its origin in nature conservation and is primarily focused on loss minimization:

Risk Management is the art of balancing risks and opportunities in order to achieve the highest overall wealth,

Sustainability tries to achieve a triple win situation, in ecologic, social and economic terms, with the smallest overall loss.

In all types of undertaking, there is the potential for events and consequences that constitute opportunities for benefit and threats to success. Risk management is therefore today recognised as being concerned with both positive and negative aspects of risk.

Sustainability is similar to the safety field, in that it is generally recognised that consequences are only negative. The policies of sustainability are therefore still focused on prevention and mitigation of harm to nature and the environment.

Liability is imposed by law on individuals and economic actors. Everybody therefore does risk management all the time, consciously and unconsciously. Every time somebody crosses a street, drives a car or signs a contract, a balancing of the risks and chances involved is performed, most of the time unconsciously.

The liability issues in sustainability are of moral nature. The degradation of the rain forests and the thinning of the ozone layer are losses of immaterial wealth due to the industrial activity of mankind, and the actors involved do not necessarily know the consequences.

Risk management is driven by liability and responsibility for third party losses that can be fought in civil courts; sustainability is driven by liabilities based on cultural values of individuals and communities for which there is no, or not yet a, jurisdiction.

Overlaps

The two concepts overlap in a number of areas, such as social, cultural and natural risks:

For social risks, factors of un-sustainability such as poverty often go hand in hand with crime, creating un-sustainable communities and risks to society.

Many cultural risks are caused by differences in culture, political systems and religion; they are fuelled by factors of un-sustainability such as a lack of tolerance. In the last few years, these risks have also been referred to as a 'clash of civilizations'.

Natural risks are multiplied by un-sustainable lifestyles; in many cases, natural events turn into natural catastrophes through changes in human behaviour. Most natural-technical disasters are caused by land-use policies, such as new towns, infrastructures and factories built in hazardous areas, such as flood plains and coastal regions. The reasons for this risky behaviour are varied and include poverty, affluence and industrial or economic development.

Uncertainty, Hazards, Risks and Risk Management

The difference between risk and uncertainty has been defined by one of the Grand Masters of Risk Management, Frank Knight, in 1921:

> 'If you don't know for sure what will happen, but you know the odds, that is risk: If you don't even know the odds, that is uncertainty.'

The beginning of the third millennium has reminded the World of the existence of uncertainty as a basis of human life, and of the cruel nature of many risks. They are sudden in their appearance, unforeseeable in their consequences – examples of the unthinkable, occurring in the most unexpected places, according to Murphy's Law that 'anything that can go wrong will go wrong'.

In the late 20th century, many risk experts had concluded that nature was the main source of catastrophes, as earthquakes, tornadoes and floods were the causes of the biggest economic losses of the recent past. This view was overthrown by 'Attack on America' on September 11, 2001, a disaster caused by people acting destructively as terrorists.

The year 2003 has then brought back some of the old hazards that have accompanied human life since its beginning, epidemics in the form of SARS and risks arising from the conquest of new frontiers, such as space technology.

In early 2003, the spacecraft 'Pioneer 10' has finally gone silent. It was launched March 2, 1972 for a mission estimated to last 21 months. Pioneer 10 first travelled through the asteroid belt and then supplied us with the first close-up pictures of Jupiter before reaching the edge of our solar system. On March 31, 1997, 25 years after its start, its mission was officially terminated. However, Pioneer 10 has continued sending signals until January 22, 2003, when it was travelling at a distance of about 8 billion miles beyond the edge of our solar system. Pioneer 10 now continues silently towards red giant Aldebaran in the constellation of Taurus, which it will reach in approximately two million years.

174

And in the same month that the unexpected success story of Pioneer 10 came to a sudden end, the space shuttle Columbia was lost during its entry into the atmosphere over Texas. The latest findings do not exclude that it was hit by one of the thousands of debris from earlier space missions, which caused damages to the edge of one of its wings that led to the complete loss of the Columbia and its crew. This was the second loss of a shuttle in about 120 missions. A loss rate of less than two percent is common in many other business areas, such as credit risks, but puts in doubt the future of this type of space technology.

The Changing Objectives of Risk Management over the Centuries

François Ewald has shown that we have gradually developed into an "insurance" society over the centuries (Ewald, 1989):

- The 19th century was greatly influenced by the ideas of social security in the work place,
- The 20th century is characterised by the development of risk-management, which, being based on the theories of risk-minimisation and loss prevention, has been strongly influenced by areas of technology, followed by investment and finally the appraisal of technological consequences,
- The 21st century, according to Ewald, will be greatly influenced by the use of the 'precautionary principle'.

It would, however, be utterly wrong to use the precautionary principle as a synonym for preserving the status quo! Modern risk management has not only broadened the scope of 'accidents' that it has to consider. It also has accepted that the importance of balancing risks versus opportunities – avoiding risks can be the biggest risk of all!

Few areas are more explicit than space to illustrate that chances and risks in space are both equally unpredictable! If the chances, including the reasons for unexpectedly successful missions, are difficult to fathom or border a miracle; equally difficult to comprehend are the risks which can materialize suddenly and without warning. But we normally only hear the disasters, not success stories, such as Pioneer 10!

To complicate matters, we have to consider that risks hardly ever travel alone, and that risks change permanently! Adaptability, resilience and redundancy thus become the key strategies in looking for solutions to prevent losses.

The 'Risk' Market

Frey (1992) identifies three main areas, namely risk assessment, risk control and risk financing.

- Risk Assessment primarily differentiates between risk perception, risk identification, risk analysis and risk evaluation.
- Risk Management and Risk Control then offers, for the risks identified, the options of ignorance, avoidance and minimisation.
- Individuals and economic actors can minimise their risks through risk financing, with the options of bearing, sharing or transferring the risk; insurance has a key role to play in the case of risk transfer.

Yet it is crucial to underline that risk assessment is not objectively determined by a 'risk' action alone, but at least as much by the anticipated profit. The 'profit of unknown risks' is often only recognised with hindsight, as in the case of Christopher Columbus and America, or Newton and the apple. In other cases, risks are consciously taken for their potential huge profits, for instance by racing rivers, big game hunters or casinos. The risk takers in the latter case act consciously, believing to have a certain degree of control on the outcome of the risk situation.

The integration of society and technology has several fundamental effects: to cut the link between risk taker and risk, and to create an inherent vulnerability of technology.

The vanishing link between responsibility and risk taker

At an individual, group and national level, both society and individuals are increasingly transferring risks. As a result, the risk taker is no longer the major risk-bearer, and thus cannot see any personal advantages in minimising the overall societal risk.

Nothing explains this better than the development of the ratio between number of civil casualties and soldiers killed in wartime:

- in the First World War, 20 soldiers were killed for every civilian.
- in World War Two, the ratio was 1 to 1.
- in the Korean War of the early 1950s, five times more civilians than soldiers were killed; and
- during the Vietnam War in the 1960s, the proportion reached 13 civilian casualties for every soldier killed!

The Titanic Syndrome

The technological risk in itself is determined by the relation 'mankind-technology', as well as by technological developments. With regard to the issue *Men and Technology*, the additional safety arising from the technological progress of 'engineering-mankind' through the progress seeking 'user-mankind' is often squashed or even reversed. The 'Titanic', the first 'unsinkable' ship in history, sank on its maiden voyage across the Atlantic due to the fact that the captain really believed it to be unsinkable. The founder had indeed announced this but had not meant it literally: the 'Titanic' was unsinkable under the assumption that the user (i.e. the captain) would behave as before. On 15 April 1912, the Titanic struck an iceberg. 1513 people died, and Lloyds of London paid out 1,400,000 pounds (in 1912 currency!) in insurance claims. Despite this, the bankruptcy of the 'White Star Line', owner of the Titanic, was inevitable.

(According to the latest technological research, "cost-consciousness" on behalf of the owner was at least as much to blame as the ice-berg itself: the cheap sheet iron used for the hull is extremely brittle at low temperatures, although people were not aware of this fact at the time. Two sister ships of the 'Titanic' provided long years of service despite their sheet iron hulls).

Modern examples of the "Titanic"-syndrome include "ABS" (anti-skid system) and 'Four-Wheel Drive'. These inventions give rise to new 'use-misuse-incentives', comparable with the 'Titanic'. The invention of the ABS system for automobiles enabled a reduction of the stopping distances and maintained full maneuverability while braking. When the system was first introduced, it was expected that the extra safety it provided would reduce the number of road accidents, and insurers consequently offered lower premiums for cars fitted with ABS. A few years later, the German motor insurers had to cancel the premium reduction for ABS, as there was no significant difference between ABS and conventional braking systems in the accident statistics. What had happened? Too many drivers had relied on the extra safety and had started taking more risks, especially in slippery conditions. The extra safety from ABS, however, would only take effect if drivers continued to drive as if they had no ABS.

The Titanic is the classic example of the syndrome of an overcompensation of the advantages of a new technology by the human operator. Captain Smith had two possible routes for crossing the Atlantic: the safer southern route or the faster, but riskier northern route. Since the Titanic was declared unsinkable, his choice was obvious. For there was another challenge: to conquer the 'Blue Ribbon' on her maiden voyage, the distinction of the fastest steamer across the Atlantic, worth a lot of publicity and a competitive advantage to attract prime passengers.

Insurance can have a similar effect as technological progress. Anybody who is insured against river flooding can build his house closer to the river, because he will be indemnified for any flood damage suffered.

The increasing complexity vulnerability through technological progress

Research into new technology, as well as technology policy, are driven or pulled by a number of factors, such as competitiveness and the ability of protection under intellectual property rights. Resilience and redundancy of technology applications are normally not among these factors.

Over the last hundred years, technologies have become more efficient and safer but also inherently more vulnerable. This vulnerability is part of uncertainty, which means that it can manifest itself on its own or be exploited through abuse and terrorism.

Hammers are a simple example to illustrate the increased complexity of new technologies; doors of vehicles can explain the increased vulnerability of new technologies.

Anybody can use a hammer without much explanation, the major risk being to hit one's own thumb. Using a personal computer – the tool equivalent of a hammer two hundred years ago – demands a considerable knowledge; the absence of this knowledge, or its abuse, can cause considerable damage or exclude people from the benefits of its use.

If the door of a vehicle opened unexpectedly, this posed a small problem in automobiles. In modern airplanes, however, losing a door in flight has led to a number of accidents and disasters, sometimes involving the total loss of aircraft and passengers. Even if best knowledge is applied to the design of doors – which it is – the inherent catastrophic risk in mobility has greatly increased in going from cars to airplane. The overall safety of using airplanes, however, is much higher than of using cars. Worldwide, an average of 1'500 people die each year in aircraft accidents; in the EU alone, 42'000 people die annually in road accidents!

The issue of risk management and technology gets more complex if we include systemic consequences of say emissions into the atmosphere, and the possible consequences for the climate. A recent study for the European Commission came up with a clear message for technology policy makers. RTD policy should orient itself on two factors hardly considered today: the perceived scale of the problem (risk perception) and the capacity to respond (risk control). Following such a policy also for existing installations and technologies should enable to

avoid solutions that fall into the panic-stricken top right-hand quarter, and promote the quest for new solutions in the safer bottom left-hand quarter.

Natural Catastrophes

In the Swiss Alps, the position of villages and churches as well as the course of
avalanches can be read from the position of the avalanche protection forests,
which have a triangular shape pointing towards the peaks. It is apparent from
these pictures that not every avalanche is an unpredictable *natural* catastrophe.
Nevertheless, occasional disasters can happen in exceptional years, causing
clashes with mankind and its technology. However, risk perception and assessment based on past events can lead to successful measures of risk control, with
regard to a particular objective.

A similar story can be told for floods. On the 1st February 1953 a breach in a
dike on the North Sea resulted in one of the worst natural catastrophes of our
time, flooding large parts of the Netherlands and drowning 2,000 people and
250,000 cattle. Fortunately, this was a one-time event. For it became a regular
occurrence, this would indicate that the risk perception of society was wrong!
Should the flood then be seen as a process whereby the sea reclaim her inherent
land?

Issues such as the "green-house effect" or the "hole in the ozone layer" could
both permanently change nature and the environment. Should they therefore be
accepted as part of the changing nature of risks, for which the best strategy is an
adaptation of our human behaviour? Or should man fight a loosing battle in trying to master nature?

A possible change in the climate is a catastrophe also resulting from mankind's modern industrial activities, and it could lead to a breakdown of the conditions that enable mankind to survive on the planet. However, other events, such
as volcanic eruptions or meteorite showers, have in the past resulted in permanent
changes to the atmosphere equating to negative effects for both the climate and
living organisms.

Living with nature may thus imply that people will have to adapt to nature's
conditions, as is suggested by an analysis of the major catastrophes around the
world (UN Department of Humanitarian Affairs, 1994) that showed the risks of
nature for man:

• Ranking disasters by damages gives the following order: floods, tropical
storms, droughts and earthquakes.

• Ranking disasters by number of lives lost shows another order: droughts, floods, epidemics, tropical storms and earthquakes.

The conclusions in a nutshell: Drought – a lack of water – kills 25% more people than floods. And nature in the form of all epidemic diseases taken together takes a larger toll than any other manifestation of nature!

The ranking of natural disasters thus greatly depends on the factors considered (human losses, economic damages, insured losses). And risk perception – understanding risks – is the key factor in focusing the objective of risk management as a strategy to increase human well-being and overall quality of life. – Sustainability.

The importance of the hidden parts of the Risk Iceberg

The real issue hidden in natural catastrophes is not a scientifically correct differentiation between "man-made" and "natural" risks, as the majority of natural catastrophes have some kind of disastrous effects on technology and vice-versa (combined natural/technological events) (Showalter et al, 1994).

In order to justify preventive measures, we have to know the cost/benefit calculation of loss prevention. An in order to compute this ratio, we have to know the total costs of disaster, what we call the 'disaster iceberg'. The visible insured costs are in most cases a small part of the total cost to the economy – but it is this total cost that we have to know.

The Health and Safety Executive in London was the first authority to study the phenomenon of the iceberg for accidents at work, and has come up with the following estimates: insured costs are between 3% and 12% of total costs!

Similar calculations have been made for other types of accidents risk, and compared to the costs that would have incurred in changes of design to prevent the disaster from happening. The near collapse of a bridge on the Swiss Gotthard-motorway, which buckled after a flood had swept away one of its pillars, could have been avoided by an increase in the initial building expenses of 1%, either by reinforcing the concrete slab, or by setting the pillar's foundation five metres deeper on solid rock. In cases such as the 'Cypress Freeway' in San Francisco, a multi-storey motorway in an area renowned for its earthquakes, which had been built by flouting the American standards on earthquake resistant constructions, the risk assessment process was ignored. The motorway collapsed during the major tremor in the 90's, burying a number of vehicles and their passengers.

Fig. 2: The Accident Iceberg

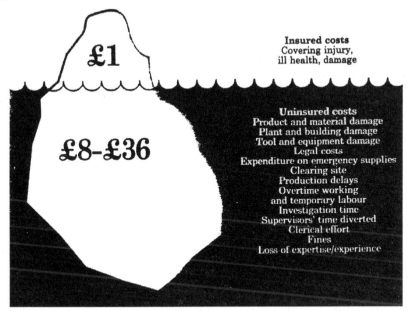

Source: HSE Health and Safety Executive, London

The relationship between costs incurred from catastrophes (i.e. insured losses as opposed to economical losses) and the expenditure, which would have been necessary for their prevention, is shown in the following figure, based on a series of well-known disasters from previous years.

Prevention is basically a question of planning ahead (or proactive management) as opposed to incurring costs end of pipe; or as *John Kletz* [3] put it: "Risk-management is not an additional coat of paint".

Taking heed from the wisdom of risk-management would in many cases bring about a slower and more sustainable way of managing the economy, e.g. the development of adaptable long-life systems built on a modular basis that can be permanently optimised in the draft and design stages.

---

3 *John Kletz* on 16 January 1989 at the ETH Zurich: "A life-cycle engineering approach" (Bernold, 1990)

Figure 3: Costs of Damage Prevention Compared to Costs Accrued from Damages

YEAR	ACCIDENT	REPAIR COSTS	PREVENTION COSTS
1976	Exploding Reactor Seveso	US$ 150m	< US$ 10,000
1981	Collapse of a footbridge at the Hyatt Regency Hotel Kansas City	US$ 90m	< US$ 1,000
1984	Union Carbide Incident Bhopal, India	>US$ 200m	<US$ 50,000
1986	Schweizerhalle Fire	US$ 60m	<US$ 100,000
1987	Flood damage to Gotthard motorway bridge Wassen	150% of initial construction costs	1% of initial construction costs

Source: Zürich Insurance Company: Catastrophe Losses – a Problem for Insurers Only? Interlaken Symposium 1987.

Conclusions

Few incentives and little know-how have been developed under the present economic or industrial policy to develop and apply strategies of sustainability or a higher resource productivity in practice. A society which 'saves the environment' by replacing ("destroy and produce new") millions of vehicles in working order when a 'cleaner' technology becomes available (e.g. unleaded petrol, lean burning engines, catalytic converters), based on industry's claim that existing engines cannot be upgraded or converted, does not act efficiently, neither with regard to its engineering development nor the environment, and acts on short-term assumptions (the proof for this was delivered by a handful of skilled Swiss mechanics converting the engines of their fleet of vintage Junkers Ju-52 aircraft to unleaded petrol).

One of the key strategies towards a more sustainable society and reduced uncertainties thus lies in a change of economic thinking towards a Service Economy. This would entail a shift in focus from efficiency to sufficiency solutions, an integration of loss prevention into economic objectives and a new focus on maintaining existing wealth in addition to creating new value.

This principles is not revolutionary – it has existed in earlier eras, as highlighted in the famous example of the Chinese village doctors.

Two thousand years ago, China already had a truly sustainable health system. Each village had its medical doctor, and each villager in good health contributed to the necessities of the MD's life. In exchange, the doctor had to look after all villagers in bad health, free of charge. This system thus had a built-in incentive that everybody in the village was better off if people were in good health – and the objective of the MD was to keep people in good health, by prevention rather than cure.

By contrast, modern health systems profit from increased levels of ill health and accidents – not good health has an economic value in Western health systems but illness !

A recent example of this alternative thinking is the policy adopted by the Dutch government after the disastrous flood of February 1st, 1953, when a third of the national territory was under water. The Dutch legislation now makes insurance policies against floods illegal, in order to make it clear that only prevention – and this means engineers, not insurers – and the upkeep of the quality of infrastructures – maintaining existing values – can avoid a repetition of the 'natural' disaster of 1953.

VI. Social and Human Responsibility

Right to Food

Michael Windfuhr und Maartje van Galen

⇒ Food scarcity is often a problem of poorness
⇒ It is imperative to focus on the rural poor

A brief overview of trends in food security and main causes of malnourishment

Current trends concerning the prevalence of hunger and malnourishment

In 1996, world leaders met in Rome at the World Food Summit and pledged to eradicate hunger. They made a commitment to cut by half the number of undernourished people in developing countries by 2015 (with 1990-1992 as the benchmark period). This objective has recently been reaffirmed when it was taken up in the first objective of the Millennium Development Goals adopted by the United Nations. Despite the rhetoric reaffirmation of this goal the political realisation is far from becoming reality. There is no doubt that world-wide enough resources (food, knowledge, money etc.) are available to overcome the problem of hunger and malnutrition. Nevertheless despite the abundance of food, over 840 million people (20 %) in the developing world today are estimated to suffer from chronic malnourishment; 11 million in the industrialised countries, 30 million in countries in transition and 799 million in the developing world.

Food consumption, in terms of kcal/person/day, is the key variable used for measuring and evaluating the evolution of the world food situation. The poverty food line is 2200 kcal a day per person. The world has made significant progress in raising food consumption per person. It increased from an average of 2360 kcal per person a day in the mid 1960's to 2800 kcal per person a day currently. In Western Europe and North America, food supplies are available at about 3500 calories a day per person whereas Africa can only supply about 2300 calories.

These statistics clearly indicate that food production is not the main cause of malnourishment.

While food availability is still a problem for some countries, the root cause of food insecurity in developing countries is believed to be the inability of people to gain access to food due to poverty. Poverty is often caused by lack of access to productive resources such as land, loans, etc. Rural areas are also often forgotten in national policy making. While bi- and multilateral development aids to the agricultural sectors were drastically reduced over the last decade, the same happened with national budget allocations for rural areas. Additionally natural disasters, such as floods, earthquakes, hurricanes, war, and civil strife leave many people without income, and consequently without food.

Who are the vulnerable groups and what are the main causes of malnourishment?

Most of the people who are malnourished are found in Asia (61 %), while sub-Saharan Africa accounts for almost a quarter (24 %). The region with the highest percentage of people who face malnourishment is in Sub-Sahara Africa and the number is declining only slowly marginally in the last years. Considering the rapid population growth in this region, this means that the total number of malnourished people in sub-Saharan Africa will increase significantly, up to 196 million people. The per capita food consumption is 2200 kcal/person/day.

The main cause contributing to the high prevalence of malnutrition is armed conflict in Sub-Saharan Africa. The frequent violent conflicts between adherent's of different warlords and politicians drive away many farming communities from their lands, destroy traditional markets. Thus causes refugees who have lost access from their usual source of food supplies. This local fighting has rarely anything to do with increasing resource scarcity; it is a struggle for power between local political leaders and often creates a destruction of available resources. One of the countries of Sub-Saharan Africa facing severe malnutrition is the Democratic Republic of Congo. The number of malnourished people increased significantly, up to 36 million. War is the main contributor to famine in the Democratic Republic of Congo. The prolonged civil war has resulted in over 2 million displaced people. Consequently, production has been interrupted and the food that is produced has mainly been distributed to armies to feed their soldiers or has been used to be traded for weapons.

Besides the impact of civil wars and droughts, several other important reasons for the prevalence of hunger and malnutrition in Africa can be identified. Many governments neglect rural population and do not invest into agricultural production. In particular, the food production mainly done by women is often ignored. Women often have no access to loans, land titles etc. In addition the rapid opening of markets, due to structural adjustment programmes, has placed African small farmers into an often difficult competition with subsidised production countries in Europe and the USA.

In India and Bangladesh are the main contributors to the high number of persons being hungry or malnourished. With 233 million people suffering from hunger, the malnutrition levels in India are among the highest in the world. The daily per capita dietary energy supply is 2430 kcal. At the same time the food production is big enough to feed the entire population. India has currently a surplus of 60 million tons of cereals, higher than the surplus production of the EU. The problem behind the two apparently contradicting figures – high level of surplus, high number of hungry – is social exclusion. The number of landless people in India alone is estimated to be around 160 million, most of them facing several forms of discriminations (as casteless, landless etc.). Giving people options or choices to earn their own income is the central challenge to overcome hunger and malnutrition in India and South-Asia as a whole region. At the same time, India has an enormous amount of undernourishment and significant many overfed people. For instance, around 50 % of the children under the age of five, living in the rural areas, are undernourished. However, in urban areas, the over nutrition problem is shooting up, thanks to the change in life style and food habits. Therefore, hunger in India is not a problem of production but of equality.

Bangladesh has 40 million people suffering from malnutrition due to similar reasons as in India. Additionally, the very densely populated country has been affected by widespread flooding for many years now, which is becoming worse through deforestation in the Himalayas and through climate change. Frequently, the poorest farmers are located in the most risky areas. While food remained available in many districts, it did not reach poor farmers, who often had lost their crops because of the last flooding.

Also, other continents have severe problems with hunger and malnutrition, even if the figures are not as striking. For example, Afghanistan and Iraq have had the greatest increases in the number of malnourished people. Afghanistan suffers from a huge rate of malnourished people (70 %). Years of insecurity and war, coupled with three successive years of severe drought, have exposed 14.9 billion people to extreme hardship. In addition to destroying crops and food supplies, warfare also disrupts the distribution of food with siege and blockade tactics. In Iraq, recent years of drought, economic sanctions, and the war against Iraq have left a large number of people to extreme hardship. 27 % of is population is malnourished which corresponds to 5.9 million people.

The high increase in malnourished people in Latin America can be explained by several factors: the economic shocks as in Argentina, the economic crisis which has hit the coffee-sectors with the lowest prices since decades or by the highest level of inequality in the distribution of national income in the world. Argentina had a dramatic increase in the number of hungry during the recent economic crisis. The number has reached 25 % of the population. Brazil has the highest amount of malnourished people in Latin America. It is estimated that between 40 and 50 million Brazilians are facing hunger and malnutrition. The majority, 35 million people, belong to the group of landless, rural poor. While Brazil has become one of the most powerful economies in the world and one of the world's largest agricultural exporters, hunger is a rural phenomenon. The positive signal is that the newly elected president, Lula da Silva, he has initiated a "zero hunger" programme and has made the ending of hunger one of his first priorities in office.

These examples make clear that hunger and malnutrition is often not caused by food shortage or lack of production, but by inhibiting access to affected people and groups to either resource to produce food on their own or to other income opportunities. Still, hunger and malnourishment is basically rural phenomenon. According to the latest study of the International Funds for Agricultural Development IFAD (from 2001), around 75 % of all poor people, suffering hunger and malnutrition live in rural areas.

What they need are responsible governments creating options for the poor groups, and an economic environment that does not discriminate against rural areas and small producers and that give countries the possibility to support and foster the development of the particularly vulnerable groups. A human rights approach to hunger and malnourishment will allow these groups to hold their own governments accountable.

Voluntary Guidelines on the Right to Adequate Food:

Reinforcing the General Demand for a Framework Legislation

The World Food Summit: five years later (*WFS:fyl*) ended in June 2002 with a very weak Final Declaration, whose sole advantage is to find in its article 19 the demand from theFood and Agriculture Organization (FAO) to develop within two years"guidelines to support Member States in their efforts to achieve the progressive realisation of the right to adequate food". Representatives of civil society organisations (CSOs) and non-governmental organisations (NGOs) expressed their collective disappointment about the lack of commitment of states but seized at the same time the opportunity to contribute in a joint process to the contents of the "Voluntary Guidelines" as "stakeholders in the work of the open-ended intergovernmental working group set up to prepare such guidelines. To make use out of the only concrete result of the WFS:fyl is even more important if recent figures of people facing hunger and malnourishment are taken into consideration. The slow pace in reducing the number of hunger were commented by the CSO/NGO community in Rome by saying "only fundamentally different policies, which are based on the dignity and livelihoo*ds of communities can end hunger"* *will lead to considerable improvements.*

Recent figures published by FAO are alarming. Without counting the progress in China, the number of people suffering from hunger and malnutrition has increased by 40 million since 1996. With the exception of China, the situation has deteriorated in most parts of the world, particularly in Africa and even in some developed countries. Income distribution has worsened everywhere, making the situation of vulnerable groups even more precarious. This is a scandal as there is no doubt that worldwide resources (food, land, seeds, knowledge, and money) are sufficient to overcome hunger and malnutrition. Even in several countries where people lack adequate food, there is enough food available and, sometimes, stocks. The lack of progress and, too often, the worsening situation should compel governments and international organizations to give priority to the fight against hunger and call for a thorough revision of development policies and hunger-reduction strategies. There can be no complacency in analyzing the causes of such an unacceptable situation.

The international human rights organisation FIAN[1], which is working on the right to food, Protest and Development Service from Germany EED, the World Alliance on Nutrition and Human Rights (WANAHR) and the Jacques Maritain Institute called in a seminar in Mühlheim/Germany to prepare a common NGO/CSO contribution to the guidelines process. The organisations present developed a common strategy and are trying to influence the process of the elaboration of the guidelines since then. A drafting committee was appointed and has written a "Joint North-South Contribution" which is open for endorsements by the civil society organisations and which is used as a background document for the process. Additionally, the drafting group produced an annex to the Joint Contribution.

What has happened since the World Food Summit: five years later

In November, the FAO Council took a formal decision to set up an Intergovernmental Working Group (IGWG) to develop the voluntary guidelines for the implementation of the right to adequate food. It was decided that the IGWG will meet three times. The first meeting was in March 2003, organized as a hearing of

---

1 If you would like to join the NGO / CSO working group please contact the International Secretariat of FIAN (Michael Windfuhr Windfuhr@fian.org or Yifang Tang tang@fian.org). FIAN can provide you with a copy of the Joint Contribution. Moreover a report is available about the first meeting of the IGWG in March 2003. The Joint Contribution is also open for further endorsements.

positions from all interested stakeholders. The NGO / CSO alliance has sent in the above mentioned Joint North-South Contribution and the Annex. A second meeting is planned for the end of October 2003 to discuss points of convergence and divergence. A third meeting in Spring 2004 shall agree to a draft of the Voluntary Guidelines. The Guidelines shall be discussed and adopted in September 2004 at the regular meeting of the FAO Committee on World Food Security.

The secretariat of the IGWG will prepare a first draft of the Guidelines set up in June 2003, based on all written submissions of stakeholders to the process and based on the findings of the first meeting of the IGWG, which took place in March 2003. The first draft will be presented to the Bureau. The Bureau will rework the draft and will then present a text for the second meeting of the IGWG at the end of October. All stakeholders will have time to comment on the first draft during August and September. The NGO /CSO alliance will coordinate their lobby work for the whole process. All interested NGOs and CSOs are encouraged to cooperate and contact the alliance (IGWG) (secretariat's address at the end).

Why is it useful to develop guidelines for the Implementation of the right to adequate food?

The Voluntary Guidelines are still far from being a legally binding document on the right to adequate food, which remains an ultimate objective for all those working towards the realisation of the right to adequate food. Still, there are considerable merits in the planned Guidelines. First, their preparation will provide the opportunity to better understand the causes of hunger and malnutrition. Second, the Guidelines will combine, in a mutually supportive way, legal instruments and procedures with development strategies and policies conducive to the realisation of the right to adequate food. This will provide a framework for a human rights based approach to specific programming to reduce hunger and malnutrition and promote nutritional well-being. Third, the Guidelines will increase the much needed coherence and consistency of governmental decisions at National and International levels as well as actions by the International Organisations in the field of food security. Finally, the Guidelines will, once adopted, be the first implemented focused document agreed by member states for one of the economic, social and cultural rights, thus leading the way to more specific guidelines for others among these rights.

The Guidelines can be used in the future as a reference for actions supporting, complementing or correcting government efforts. They will also be of use when assessing decisions of governments, International Organisations and other stakeholders. The CSOs will be able to persuade various actors to meet their obligations or responsibilities and they will make observations when non-compliance with the Guidelines affects the vulnerability of food insecure individuals and targeted groups in all countries. The basic problem concerning the missing implementation of the right to food at the national level is primarily not inadequate supply of food but insufficient income and lacking access to productive resources of those who are hungry. In many countries, the government has not identified the socio-economically vulnerable and food insecure groups, and therefore has no policies in place to address their problems, and hesitates to challenge vested interests or to adopt unpopular policies. Furthermore, in many countries, governmental dilemma is between providing the urban poor with food at low prices and securing sufficient income for the indigenous communities.

Assessment of vulnerability and adaptation of legal framework

While vulnerable groups are many and differ from one country to another, primary attention must be given to the fact that still close to 80% of the poor suffering from hunger live in rural areas. It is imperative to focus on the rural poor. From a developmental perspective, the only viable and sustainable solution is to give small farmers the possibility to make a living from their activity. If successful, this could slow down rural-urban migrations that lead to unmanageable cities and induce sustainable growth for the national economy. In addition, it is imperative to end discrimination against rural women regarding access to land and loans. Attention must, of course, also be given to the many other vulnerable groups.

Given the diversity of situations between countries and the variety of causes affecting the vulnerable groups within a country, each country will have to adapt the policies and legal framework recommended in the Guidelines to its specific situation. The Guidelines must therefore be drafted in sufficiently general language. Each country should be strongly invited to design its own strategy to fight against hunger and malnutrition, having in mind elements for National strategies. For example, a national strategy that will have political, institutional, legal and economic dimensions based on an accurate and comprehensive analysis of the causes of hunger and malnutrition at the National level and for each vulnerable group. Indeed, a country can design valid policies and set priorities only if they are based on a clear understanding of its agricultural, environmental, social, and economic situation. In developing appropriate legislation, a country needs to take into account its institutional, financial and human ability to implement it.

External and International Obligations

Besides to the national responsibility of the governments, the NGOs and CSOs see the need to address international circumstances. They are increasingly influencing the capacity and ability of states to implement economic, social and cultural rights. From a civil society perspective, countries should recover the necessary policy space to conduct their fight against hunger and to be able to implement fully its obligations under the right to adequate food as well as other human rights. Moreover, the Guidelines should address the impact that the national policy of a given country or group of countries may have on the enjoyment of the right to adequate food in other countries and should recognize the fact that countries have also external obligations under the right to adequate food. For instance, subsidized food exports, whether direct or indirect, may deprive small farmers in other countries of their rights and should therefore be discontinued. Members of International Organisations, governments should secure that principles, conventions, norms and policies developed in the various institutions with respect to the right to adequate food are implemented in a coherent and consistent manner.

Social Contract between the Generations: Work & Pension

Bert Rürup and Jochen G. Jagob

⇒ Poeple need to have a longer working life span.
⇒ A public pension-system is the basis for protecting the older generation from poverty.

1. Introduction

During the last decades there is a common development in every OECD country observable that an increasing share of Gross Domestic Product (GDP) is spent on their national social systems. At the same time the East European countries were faced with a drastic political and economic change. The economic and political transition in these countries had substantial effects on social policy. Social policy received a lot of attention by economists and politicians in those countries. Social insurances include basically a system providing an income after retiring from the labour force, insurance compensating income losses due to unemployment and a system insuring against the costs connected with illnesses and their medical care. Since the national labour market regulations and the national health systems differ to a great extent in their institutional settings, a comparison on a general level is rather difficult. We will therefore restrict our investigation on pension policy because in the opposite to the unemployment insurance and the health system the initial state of the pension systems as well as the problems connected with them are quite the same in the OECD and the transitional countries of East Europe.[1]

But pension policy can not be examined in an isolated way. The overall economic situation as well as in particular the situation on the labour market has to be considered as well. Given the differences in the national pension systems and their specific reactions to the circumstances we will show a general way out of the pension crisis which makes the system sustainable in the future.

---

1 Our further restriction to the OECD and the East European countries does not mean that there is no need for social policy in the underdeveloped countries. But there is a substantial difference in the overall economic performance which leads to a completely different situation. Social policy in underdeveloped countries therefore needs an overall economic development.

As a consequence we will show the general needs for reforms. We will examine the adequate alternatives to make the pension systems sustainable. Finally we will summarize and conclude our results.

2. The Status Quo

2.1 The different pension systems

Every pension system serves to transfer income from a period when people earn a labour income to a period when people retire from the labour force. It is designed to provide an income when people due to their age do not participate in the working life anymore. The income transfer from one period to the other can be financed in two different ways.

The majority of public pension systems is financed by a PAY-AS-YOU-GO (PAYG) system. In a PAYG system the revenues are a part of the labour income in the economy by levying a contribution that has to be paid by the members of the labour force. The contributions of the currently working generation are immediately used to pay the benefits to the currently retired generation. In a PAYG system the revenues therefore always equal the expenditure, i.e. the budget is balanced in every period. From a theoretical point of view a deficit does not occur in such a system.[2] The equality of revenues and expenditure is achieved by either an increasing contribution rate or decreasing benefits. – As a compensation for the contribution payments during the working period each generation is entitled to receive a benefit during their retirement. This transaction produces an implicit debt since current contributions are connected with a benefit payment in the future. Thus a liability is created. An implicit debt is not compared to an explicit debt not covered by any real assets nor is any government bonds issued. The expected benefit which is solely an entitlement to receive part of the future labour income of the domestic economy is therefore not explicitly defined.

Another possibility to transfer income from the working period to the period when people retire is a fully funded system. Compared to a PAYG system a fully funded system does not use the labour income as the main source to finance old age income. Here the contributions of each employee are invested on the capital market and accumulated to an individual capital stock over the entire working pe-

---

2 In fact this is a simplification of the reality. In every PAYG pension system transfers from the national budget are a part of the revenues to cover existing gaps between the contributions and the benefits.

riod. The benefits during retirement are paid out of the capital stock plus its interest.

In a PAYG financed system no capital stock is accumulated. The benefits are merely a not specified liability as a compensation for paying the benefits of the parent generation. The lack of funding in a PAYG system means that the benefits are merely based on the payments of the current employees. A implicit public debt is created by the accrued pension liabilities which can be amortised more or less arbitrarily. The implicit debt is not reported in any official statistics though it definitively has the effect of an intergenerational redistribution like the explicit debt. The redistribution appears in the way that the burden of paying back the debt is shifted from the currently living generations to the future generations.

Besides the differentiation into a PAYG and a fully funded system pension systems are also distinguished according to the social policy aims that are desired to achieve. Avoiding poverty in the old age is one of the possible goals of a public pension system.

A redistributive pension system usually fits best to achieve the goal of avoiding poverty in the old age. Systems like that are usually entirely taxes financed and the benefit level is fixed as a certain percentage of the average wage income which usually corresponds to the poverty line defined by the OECD and other international organisations. Any needs beyond the basic necessities have to be financed by private savings. Since these systems are usually tax financed the burden of financing the revenues is therefore in contrast to a contribution based system spread over the entire population in an economy. Even pensioners themselves may finance their own pension if consumption taxes like the value added tax is used for the revenues. Since under determined conditions everybody is entitled to receive such a pension which has a commonly fixed benefit level independently of how much was individually contributed to the system there is tending to be a redistribution from high income earners to low income earners.[3] Since redistribution in such a system happens within one generation it is denoted as intergenerational redistribution.[4]

---

3 The entitlement to receive such a pension usually depends on the nationality or the time of residence in the country. Admittedly in most cases the time of employment in the country and therefore the time of contributing is taken into account when giving the allowance to the benefit.

4 Such defined benefit systems with a benefit level that attempts to avoid poverty are often connected with the name of Lord Beveridge who elaborated a system for social security which is based on the principle of avoiding poverty. But even in systems like that benefits are often tied to a minimum time of employment. Such systems can be found in countries like the United Kingdom, Denmark, and the Netherlands.

The opposite aim of pension policy is to preserve the same standard of living for the people during retirement than before. Such system is usually financed by contributions.[5] Though there is no system in the world which is entirely financed by contribution payments because transfer payments of the general public budget are more or less common in every pension system, the contributions have a major importance since the individual benefits depend crucially on them. Compared to a pension system of a fully funded type with a fixed benefit in a Bismarck system the benefits are individually assessed based on the individual contribution payments during the working period. Since everybody who earns a high labour income has to pay higher contributions than some body with a lower wage income they are the ones who receive a higher pension benefit. Any employee therefore stays comparatively on the same income level during retirement than before. Since the system is finance by contributions which equal a certain percentage of the labour income[6] the revenues depend to a high degree on the situation on the labour market. A situation with a high unemployment causes lower revenues for the system which causes in the short run financial instability.

A PAYG financed system can be differentiated in a system that aims at avoiding poverty in the old age and a system that keeps people during retirement comparatively on the same standard of living as in the working period. In spite of these differences in the institutional settings every PAYG system has in common that its main source of revenues is the factor labour and the income produced by it. It therefore depends to a high degree on the stability and the productiveness of the domestic labour income. The financial stability needs an appropriate relation between the number of pensioners and the number of employees. Any distortion of this relation results in a shift of the burden between different generations. The demographic change like every OECD member country is faced to do have the effect of such a shift of the burden. The impact of these changing conditions on the PAYG financed pension system is investigated in greater detail within the next section.

---

5 A pension policy that is organised in this way is commonly connected with the name Bismarck. Though Bismarck originally installed a fully funded system a PAYG system with a tight relationship between individual contribution payments and the benefits is referred to be of Bismarckian type

6 Note that in a defined contribution system the contribution rate is fixed over the time whereas in defined benefit system the contribution rate might vary. Though it is usually constant within one period.

2.2 The Problems concerning the Pension Systems

There are two main reasons that influence the relation between the working generation and the retired generation, namely the demographic change and early retirement. Their effect on the public pension systems are investigated in the following sections in order to receive the results which serve as the basis for the necessary measures that have to be undertaken.

2.2.1 The demographic change

Comparing the demographic development in the different OECD countries and the countries in Eastern Europe a similar tendency in the future population structure is observable. The development can be summarised in the way that there will be an increase of the average life expectancy and decreasing birth rates. The result is an ageing society on the average. There is an immediate effect on the PAYG financed public pension systems due to this development.

Every PAYG system has the property that its revenues depend on the number of contributors. Since contributions as well as taxes to the main degree are levied on the labour income the revenues depend on the number of employees in the economy. The demographic effect of low birth rates results in a shrinking number of the future working generations and therefore tends to decrease revenues of the system. Simultaneously the expenditure in PAYG systems depends on the number of retired persons. Based on the demographic development an increasing number of pensioners could be expected in the future. Two reasons are responsible for this development. On the one hand the baby boom generations will retire within the next decades and on the other hand will the increasing live expectancy have the effect of a longer period of drawing benefits. Both effects result in a higher number of pensioners.

Observing both the revenue side and the expenditure side of a PAYG system it has to be concluded as Rürup and Liedtke [1998] point out that the financial stability of such a pension system depends to a high degree on the relation between employees and pensioners. In Germany there are currently 30 mill employees who equal basically the number of contributors and 20 mill retired people. Supposing the current demographic development the situation will be even worse because of a lower number of employees due to the low fertility rates and a rising number of pensioners due to the increasing life expectancy and the baby boomers who are about to retire within a foreseeable period of time.

The consequence of a development like this will be either increasing contributions in order to maintain the fixed benefit level or decreasing benefits if the contributions are kept at a constant level.

Both increasing contributions and decreasing benefits are politically and economically hard to handle. Especially a raising contribution rate does have the effect of increasing the costs of labour as a factor of production in the economy. An increasing contribution rate with fixed benefit level results in labour market distortions which has the effect of a shrinking labour market or, evading the contributions by participating in the shadow economy. Shrinking benefits are undesirable because it could harm the aim of protection of the pension system by violating the confidence in the system. Besides that the political feasibility of decreasing benefits becomes in an ageing society more and more impossible. The raising number of old people results in a higher political power of this age group. Any government that attempts to decrease benefits therefore jeopardises to get re-elected.

Thus the public pension systems are faced to the problem of ageing. The ageing of the society causes costs for the public pension systems if the current policy is going to be maintained. The burden of ageing is presently borne to a main part by the working generations. But no attempt pension reform is able to abolish the costs connected with ageing it is needed to modify the system in a way that the costs caused by the demographic change is not burdened in an unilaterally way on the factor labour.

Considering the present situation of the public pension systems in the member countries of the OECD it is possible to conclude that the demographic change will have the effect of shifting the burden which almost entirely will affect the currently young and future generations. Those generations have to bear the highest share of the costs connected with ageing of the society. The extent of the shift of the burden can be measured by the method of generational accounting, internal rates of return, and the implicit tax rate of the public pension system. All of these measures indicate a substantial burden for the future generations and that this burden is an immediate effect of the demographic change in the industrialized countries. This is immediately evident by the internal rate of return which is as Aaron [1966] and Samuelson [1958] showed the product of the wage growth and the population growth rate i.e. the overall economic growth. In a situation with low fertility rates a shrinking internal rate of return is a direct effect as long as it is not compensated by a higher wage growth which could not be observed in any of the OECD member countries. Moreover over the last decades the internal rate of return used to have a value that was always substantially below the capital market interest rate. Based on the demographic development the spread between both of these values will be enlarged in the future. Consequently people would be better off if they had invested the same amount of money as savings on the capital market.

A similar result is achieved by the method of generational accounting. A generational account is the present value of the deviation of the sum of the average contribution payment of one generation and the average sum of benefits received by the same generation. A positive value of this deviation indicates that a member of the observed generation pays more contributions as she receives as benefits in return. A generational account with the value of zew that contributions equal the benefits. Calculations based on this method of Auerbach, Kotlikoff, Leibfritz [1999] and the European Commission [1999] show unambiguously that the generational account of the future generations has a higher value than for the currently living generations. Consequently the future generations will be faced with a higher burden due to the demographic development which has to be borne by the labour income of these generations.

The implicit tax rate is the relation between the deviation of average contributions and benefits and the lifetime income. The implicit tax rate of the German pension system was calculated by Thum and von Weizsäcker [2000]. It shows a raising tax rate for the currently young and the future generations. An implicit tax is compared to an usual tax a result of the deviation of the contributions and the benefits which increases in the future due to the demographic change. The factor labour will therefore be burdened to a higher extent.

The influence of pension policy on the demographic development and therefore on the process of ageing is though limited. It is rather the responsibility of the family and education policy to create incentives which could have an effect on fertility behaviour. In detail the decision to bear and raise a child has to be made more attractive in the society by reducing the costs connected with it. The costs of children can be divided into direct and indirect costs. While the direct costs can be handled by a transfer system the indirect costs like wage loss during the education period or difficulties to re-enter into working life to comparable conditions to the situation before leaving it have to be reduced. Taking the education period into account for the benefit level is just one way to achieve this aim within the pension system itself. Beyond this problem pension policy is limited to achieve a situation where the burden of ageing is redistributed in a fair way on every generation involved. Pension policy therefore has to lower the burden on the factor labour in the long run without cutting benefits to a too large extent. A further burdening of the factor labour will have a negative effect on employment and economic growth.

The burdening of labour by the pension systems has a significant effect on the interaction between the labour market and the pension system. Within the last year a tendency towards early retirement could be observed in nearly all OECD countries. This tendency appears in the way that people prefer to leave the labour force by retirement rather than working beyond the earliest possible age of retirement. There might be many reasons for a development like this but an important effect is definitely the burdening of labour. The effect of this development is a shift of the relation between the number of employees and the pensioners in a direction which worsens the financial situation of the PAYG pension systems. These effects will therefore be investigated to a further extent in the next section.

2.2.2 Early Retirement

Though in every pension system exists a regular age of retirement over the last years it could be observed that people tend to leave the labour force at a significantly earlier age. Within the same period of time the labour market participation of the age cohorts between age 60 and 64 diminished in most of the OECD countries. Compared to every other age cohort older than 20 this age cohort has the lowest labour market participation rate at all. If any action in order to increase the actual average retirement age is planned it has to be considered that there are different channels for retiring like the disability retirement for example.

Disability retirement is in fact a frequently used way to retire early but it is generally only granted due to major health problems that make a further employment impossible or at least it just could be continued to a limited extent. Since disability is usually not a problem that only affects older age cohorts it might serve as an explanation for the deviation between the actual and the legal age of retirement but it is no explanation for the low labour market participation of the age cohort 60-64.[7]

Besides disability most pension systems have the possibility of retiring before the regular age of retirement. Though it is usually connected with a proportional reduction of the benefits. But with this reduction of the benefits an incentive to leave the labour force to an earlier time as predetermined often remains. As it was pointed out by many economists before the concept of the marginal fairness has a prominent importance at the individual decision about retirement. The marginal fairness reflects the additional benefits resulting from an additional period of work and therefore an additional period of paying contributions. A higher degree of marginal fairness in a pension system therefore reduces the incentive to

---

7 The risk of becoming disabled is usually spread more or less randomly over every age cohort. This is at least for most injuries and accidents which cause disability the case. Whereas other health problems like chronic diseases which are to a high extent an effect of the working conditions might appear with a higher probability the older a worker is.

retire early. Given this coherence between marginal fairness and early retirement a system of Beverdige type with fixed benefits which are independent of the contributions has a rather low degree of marginal fairness and therefore lower incentives to stay longer in the labour force.[8]

The incentives given by the institutional settings are concerning the problem of early retirement just one part though an important one. As Siddiqui [1997] shows in an empirical investigation for the German pension system did the permission to retire earlier have the effect of the development now observed. The demographic change worsens the situation insofar as it has an effect on the distribution of the burden between generations – measured by the implicit tax rate or the internal rate of return – and therefore is a deviation from the principal of actuarial fairness[9] of the system.[10] In the absence of the principle of actuarial fairness it is rational for the individuals trying to gain as much benefits as possible. One way to achieve this situation is by early retirement. A surprising result has been discovered by Riphahn and Schmidt [1999] and Siddiqui [1997]. According to their investigations a high unemployment has no effect on the retirement behaviour. In fact the opposite is the case since people expect a lower income they try to stay in employment. But this is only a relevant decision for those people who have a job. For the unemployed part of the population leaving the labour force and therefore leaving unemployment by retiring early is an attractive alternative since especially for the unemployed of the older generations the probabil-

---

8 Another example for a system with a low degree of marginal fairness is the Austrian pension system . Though it is a system financed by contributions the individual level of benefits is calculated on the basis of the 15 years with the highest labour income. Additionally working between the age of 60 and 65 had a low weight according to any increase of the benefits. The effect is a low incentive to work beyond the age of 60.

9 An actuarial fair insurance is defined as a system where the present value of the sum of contributions equals the expected present value of the sum of the benefits. In a pension system no shift of a burden between generations takes place if the principal of actuarial fairness is fulfilled An actuarial fair system is equivalent with a system that has Generational Accounts and an implicit tax rate with a value of nil and an internal rate of return that equals the capital market interest rate. To be more precisely this means actuarial fairness on average. an even harder requirement is the condition of individual fairness. In this case actuarial fairness has to be fulfilled on the individual level. Such a system does have neither any intergenerational nor any intergenerational redistribution.

10 Note that an actuarial fair PAYG system is only given if the internal rate of return equals the capital market interest rate. This case would be incidentally rather than a result of the system . Since every payment is discounted with the capital market interest rate by the individuals the present value of the contributions equals the present value of the benefits in this case.

ity of finding a new job is usually very low. This is exactly the reason why there is no coherency between early retirement and the entry of young generations in the working life. Moreover corporate early retirement actions are not used in order to exchange older employees against young ones. It is in most of the cases solely a way to decrease employment in the firm and simultaneously avoiding to increase official unemployment. Early retirement is therefore only a way to shift the costs of unemployment from the unemployment insurance to the pension system.

The development of early retirement is a problem that creates costs for the society. On the one hand it causes costs for the pension system by altering the relation between contributors and pensioners to a less favourable one because due to the longer time of receiving benefits the expenditure is rising. This effect will be even larger with increasing longevity. Additionally revenues are decreased by early retirement as well. Both effects destabilise the financial situation of the system. Another effect of early retirement is as Herbertsson and Orszag [2001] show a welfare effect on the macroeconomic level by reducing the potential output of the economy.[11]

The financial stability of the present pension systems is hardly maintainable. The instability is mainly caused by the demographic change which results in a substantial intergenerational redistribution. Besides this rather exogenous effect instability of the pension system is also caused to same extent by the demographic development but to the main part by the institutional settings of the system which permit early retirement. A reform of the pension systems in order to recreate financial stability is more than necessary. The options for such reforms are examined in the following chapter at greater detail.

3. The Reform Options for Sustainability

All the pension systems as they are currently organised are faced to a substantial socioeconomic change that influences the stability of the systems. A sustainable modification of the pension systems therefore has to be an appropriate reaction on these changes. Sustainability of the pension systems means that it has to fulfil

---

11 Herbertsson and Orszag [2001] defined the potential output of the economy as the output which would result if all factors of production had been exploited. Early retirement reduces the amount of labour available for production in the economy. The consequence is a lower GDP as possible if the full amount of labour in the economy had been used. Since these are costs which are not real costs but more or less fictitious ones these are often referred as opportunity costs.

the needs of the population by ensuring transparency and a appropriate relation between contributions and benefits.

But every reform option is limited by its feasibility. A reform of the pension system therefore has to take the status quo into consideration and it has to minimise the costs of the transition to a new system. A reform should be divided into two parts: In the first part some crucial parameters in the present system which cause instability have to be adjusted in order to improve the system towards sustainability. Beyond that a second much more fundamental reform has to be done. This means at least to a certain extent that the system itself has to be changed as well.

3.1 Parametric Reform Options

The debate about the reform of the PAYG financed pension system usually concentrated to a main part on a transition to a fully funded system Especially Feldstein [1996] did favour a privatised fully funded system. Since in a fully funded system everybody accumulates a capital stock during the working life which is used to finance the benefits during retirement such a system is by far less affected by the demographic change. Though such a policy advice should always be handled with care because there are serious doubts concerning the feasibility of a transition. A switch from a PAYG financed to a fully funded system might have the effect that the problems caused by the demographic development could be solved but it also might create new problems. Most of the arguments about this consideration has been pointed out by Orszag and Stieglitz [1999], Barr [2000] , Breyer [2001] and Persson [2000].

The obvious reason why such a transition is impossible are the transition costs because a compensation for those who have accumulated entitlements for a PAYG financed pension by their contributions and have not saved a sufficient amount for retirement on their own. The working generations during transition are therefore faced to a double burden of financing their own pensions and compensation the previous generations.[12] Theoretical investigations by Breyer [1989], Brunner [1996], Brunner [1994] and Fenge [1995] shows that such a transition is not possible without making at least one generation worse off. Additionally each empirical investigation about this subject show that there will always be costs connected with such a transition. The costs connected with such a

---

12 The term double burden does not mean that these generations have to pay an amount which is twice as high as without a transition. It barely reflects the situation that they have to pay for their own pension and have to pay the compensation payments.

transition can therefore be regarded as sure whereas the benefits of such a transition are more or less uncertain.

The arguments for a public provision of a pension system from an economic perspective have been examined by Diamond [1977]. The possibility for redistribution and to spread risks upon a great number of people is the main reason for government intervention.[13] The effect of risk diversion is reducing the costs which are given by an individual risk like the risk of longevity by distributing the costs of the risk among the entire society. This reduces the individual costs of the risk if it follows a random distribution within the society.[14] In a pension system the main risks which are insured are the risks of poverty in the old age and the risk of longevity. While the risk of poverty in the old age is a typical problem that can be solved by intergenerational redistribution the risk of longevity is a typical insurance problem. It can be both solved in a PAYG and fully funded system as long as longevity and life expectancy respectively is predictable. If longevity is not predictable and connected with a high degree of uncertainty a fully funded system is of limited effectiveness. Every pension system therefore needs to have a PAYG financed pillar.

In order to achieve sustainability in the pension system in first step the PAYG systems need to be modified by parametric reforms in a way as the European Commission [2001] pointed it out:

"The essential feature of parametric reforms is that they aim to maintain the basic structure of the existing system while attempting, through changes in parameters, to influence the costs, financing or incentive structures of the scheme in order to adjust it to foreseen circumstances."

Though modifications in the PAYG financed pension schemes are necessary in order to adjust them to the changing circumstances but their basic structure has to be remained. A first action that has to be done is a reduction of early retirement.[15] One possibility to achieve this situation is a gradual increase of the regular age of retirement in order to increase the actual age of retirement. The

---

13 Especially redistribution depends on a mandatory public system. From an economic point of view redistribution on a voluntary basis is not feasible.

14 A random distribution in this context means that everybody is affected by the risk with a probability greater than nil. In the case of longevity this means that the realisation of living longer than expected could happen to everyone even though the probability of realisation is individually very small.

15 In this context disability retirement is more or less excluded from the consideration. Avoiding disability retirement has to be done by improving the health of the employees and their working conditions. But all of these measures could be influenced by the institutional settings of the pension system itself just to a limited extent.

changing of this parameter would have a positive effect on the pension system concerning financial stability. As it was described above demographic change consists of low fertility rates and an increasing life expectancy. Since a higher retirement age shortens the period of benefit receipt the effect of increasing longevity would be outweighed. Even if such a policy is not very popular within the population it is indeed a very necessary one. Since people tend to live longer in a quite good physical shape maintaining the present benefit level would result in a windfall profit for the pensioners as they receive more pension payments that have to be paid be the contributions of the working generations. Increasing the retirement age though has to be connected with a permanent education of the labour force because the demographic change does not only have the effect of a rising fraction of pensioners in the total population but due to the shrinking fraction of young people a diminishing labour supply as well. The shrinking of the labour force can only be stopped by keeping older people in employment. To achieve this human capital has to be kept on an up to date level. Exactly this is the reason why permanent education gets a important role in this context. In order to keep older workers in employment several alternatives are thinkable. One way to achieve this situation is by a general increase of the retirement age. This would have the effect that people have a longer full time working period than they have nowadays.

Another way to offset the decreasing number of the labour force which would result as a consequence of the low fertility is by part time working especially for older workers.[16] Both alternatives have to be complemented by a continued education of the labour force. Additionally it is also thinkable to create the possibility for pensioners to work besides receiving their old age benefits. This is on the one hand a possibility to employ older people and on the other hand pensioners have a further disposable income besides their old age benefits and a possibly available capital stock. This kind of labour income could be relieved from taxation up to a certain extent.[17] But it should be considered that just a low income employment which does not exceed a certain weekly work time and a certain income level should be tax free.Corresponding to three pillar system of old age security developed by the Worldbank [1994] consisting of a PAYG pillar, a fully funded pillar, and private savings this is often regarded as the fourth pillar of old age security.

---

16 Though it has to be mentioned that part time working was successfull in some countries it failed to be an effective policy tool in other countries like Germany and Austria. But keeping in mind that the labour force is going to shrink it could be an option to offset the diminishing number of workers.

17 The relieve of taxation is necessary in order to create the incentives for employment of the older workers.

But the attempt to increase the actual retirement age by a higher regular age of retirement has to be accompanied by further actions. As Siddiqui [1997] pointed out for the retirement behaviour in Germany admitting to retire before the regular age had the effect that it was used as a possibility to leave the labour force early. In 1992 a pension reform was accomplished in Germany which introduced reductions of the benefits by 0.3 percent per month of retiring earlier than at the regular age of 65. This modification had according to Siddiqui [1997] the effect that people decided to retire later than before the reform. Though the average retirement is still substantially below the regular age. But it is not the possibility of a flexible retirement decision on its own that creates problems since it leaves the decision about retirement to the individuals according to their preferences. The problem is rather to have a system that fulfils the principal of marginal fairness. Every individual that decides to retire at an age before the regular age of retirement has to be faced to the situation that this decision will reduce its benefits to a noticeable extent. If the reductions fulfil the property of marginal fairness the individuals will decide to retire at an age close to the regular age of retirement.

But marginal fairness is not the only modification that is necessary. As the European Commission [2001] pointed out a closer attention should be directed towards the gross replacement rates within the system. Strengthening marginal fairness is connected with the desired effect of an increasing actual age of retirement which raises revenues on the one hand and decreases expenditure on the other hand. But this effect is as Breyer and Kifmann [2001] show just of short term nature. Since people live longer and have to pay more contributions they would receive corresponding to current law higher benefits entitlements. The effect of an increasing actual retirement age by higher reductions of the benefits would be a higher expenditure than before the modification with a certain time lag. It is therefore necessary to adjust the benefits to the new rules.

As first step of reforming the pension systems which aims at a sustainability of the systems it could be concluded that a complete transition from a PAYG to a fully funded system is not a feasible solution since its transition costs would be to high. The aims of social policy which are income redistribution and risk sharing can be considered as an additional argument for a PAYG financed system. Nevertheless such a system has to cope with the changing social and economic circumstances. An adjustment to these changing conditions is therefore more than necessary. A lot of efforts have to be made in order to reduce early retirement which endangers financial stability and is to a large extent a self induced problem. The strengthening of marginal fairness as well as an adjustment of the benefits with a corresponding benefit calculation are the parametric reforms in a

PAYG system which have to be considered as necessary. In the next section reforms that are more fundamental will be examined.

3.2 Systemic Reform Options

The existence of risk is generally considered as the source of insurance demand. Any insurance is based on the principle that individual risk is shared by a larger number of people. A mandatory insurance manages to distribute individual risk on a very large number of people.[18] The individual costs of the risk are thus minimised. As already described above there is either the possibility to finance a pension system by a PAYG or a fully funded system. Both systems have different properties concerning their ability to cope with different kinds of risk.

Both systems have specific advantages and disadvantages. Since a PAYG system relies to the stability and the productiveness of the labour income its advantages are:

- its high immunity against inflation
- its flexibility according to an implementation and expansion[19]
- its availability to use it as a labour market or redistributive tool and
- its merits towards female as it does neglect the higher life expectancy of women compared to men.

The disadvantages are its sensitivity to employment and the demographic change expressed by the relation between contributors and pensioners. In an ageing society young and future generations are left worse off.

Compared to a PAYG system a fully funded system is based on the stability and effectiveness of the national and international capital income and respectively the capital market. But in contrast to a PAYG system which barely relies on the domestic production in a fully funded system contributions can be invested on international capital markets. A fully funded system is therefore independent of the domestic production and employment. Demographic change affects it by far less than a PAYG financed system.

The most important disadvantages of a fully funded system is its sensitivity against national and international capital market risks and its rather low possibilities to integrate social elements like redistribution.

---

18 In an extreme case it is the whole population of one country.
19 Especially after reunification in Germany the former West German system could be expanded to East Germany very easily

Given the specific advantages and the disadvantages of both systems and their different reactions to different economic and demographic circumstances a higher degree of protection could be achieved by using both systems to finance the benefits. Especially Merton [1983] came to the result that using both factors of production as the mean to finance the pension system could result in a positive welfare effect and therefore is an improvement for everybody. Reforming a pension system with the aim of establishing a sustainable system that is able to cope with different kind of risks should be based on a the domestic labour income i.e. a PAYG system on the one hand and on the national and international capital income i.e. a fully funded system on the other hand. Such a mix of the systems enables to pool the specific risks of each system and outweigh them against each other.

A further argument for a mixed system was pointed out by Sinn [1999a, 1999b]. This argument is based on the demographic effect of decreasing fertility rates and the substitutability of labour and capital. To summarise in a world were less human capital is available for the production process it has to be substituted by real capital. A funded pension system is a possibility to achieve this situation. Moreover endowing the factor labour with more capital has the effect that labour productivity will rise. An higher economic growth will be the result.

If sustainability is the policy goal of a pension reform a mix between a fully funded and a PAYG financed pension system is the adequate solution. By funding public pension to a certain extent a further effect can be achieved. Since fully funding fulfils the principle of actuarial fairness to a higher degree than a PAYG financed system it increases marginal fairness of the system. The distortions on the labour market – as reflected by early retirement – could therefore be reduced. Any attempt to move into a direction of a higher degree of actuarial fairness is from an economic perspective always an improvement to the previous situation. This is as Lindbeck and Persson [2001] pointed out one policy goal that should be achieved.

Last but not least the question how the funding should be organised has to be answered. There are three possibilities.

A first solution is voluntary savings in order to accumulate the necessary wealth which serves as an old age income. These savings could be supported by government transfers for exactly this purpose. The individuals have the opportunity to decide among different suppliers of annuities or other savings which have the property to serve as an old age income. All the suppliers are monitored by the government and therefore have a high degree of transparency. Such a system corresponds more or less with the pension reform of the year 2000 in Germany.

A second possibility of funding is by mandatory savings as they exist in the Swedish system. In such a system a constant proportion of the contributions are invested in pension funds which are privately organised but monitored by the government. It has to be possible to change the pension fund when it is desired.

A third possibility of a funded pillar in the pension system is by a corporate pension system. In order to avoid disincentives in such a system it is necessary that the savings have to be portable in way that the employees do not loose their entitlements when they change the employer. Otherwise such a system would hinder labour mobility and an inefficient lock in effect would be the consequence. A system where corporate pensions have a high importance for the pension system was established in Switzerland.

Finally it is very difficult if not even impossible to find an optimal solution concerning the organisation of funding the public pension system. It is necessary that the funded element fits to the given PAYG financed system and has the property to fill the gap that is caused by the demographic change. A further effect of funding the pension system is to reduce the costs of financing and therefore discharge the factor labour by lower contributions than in a purely PAYG financed system. The reduction of the costs for the factor labour has positive effects on the labour market.

Besides the parametric reforms described in the previous section a further systemic reforms is needed. Funding is necessary in order to substitute the missing human capital by real capital to finance old age income. Besides that a hybrid system works in a way to reduce risks of each system. The decision about how to organise the funded system depends on the existing PAYG financed system. The aim of funding is to close the gap caused by the demographic change. In general there is no one best way to achieve this situation. There are rather diverse attempts which have a similar result.

Conclusions

After the World Bank [1994] published their report about reforming public pension systems a three pillar system became an appealing solution. The three pillar system consists of a PAYG financed first pillar, a mandatory funded pillar and private savings that could be supported by transfers of the government. Though the basic idea of mixing the pension system by a PAYG financed part and a funded part is from an economic perspective very convincing some remarks have to be made.

The design of a pension reform always has to start with the initial situation. Given a certain situation a concept have to be designed that ensures the financial stability of the system in the long run and that can be achieved at the lowest costs possible. Most pension systems in Europe and North America depend on a

PAYG financed pillar because it is a justification for public provision. It enables to redistribute income and share risk inter- and intragenerationally. The funded pillar has to close the gap which are produced by the demographic change within the PAYG pillar but the PAYG system has to be preserved in its main structure. Besides this transparency has an important role in every effort to reform the pension system. One way to achieve such a transparency in the PAYG system is by the notional accounts where the entitlements accumulated are accounted as it was described by Feldstein [2001].

Besides the systemic reform of funding the system to a certain extent parametric reforms are necessary as well. The main effort is to adjust the actual age of retirement to the regular age. There are two ways which have to go hand in hand in order to achieve this situation. On the one hand the regular age of retirement has to be raised and on the other hand the existing structure has to be adjusted to the new circumstances. Last but not least it can be concluded that a sustainable pension system is a transparent pension system and a system that avoids disincentives for the individual decision making. Beyond that a system has to be flexible enough to cope with the socio-economic change without creating a financial instability.

Society and Ethos

Hans Küng

$\Rightarrow$ Responsibility of the single person
$\Rightarrow$ Changing the Ethic-value-system

Future opportunities for Humans and the Environment Perspectives of a Sustainable Development

One Ethic in One World – Ethical rationale for sustainable development

1. The limits of pure reason

(1) Pure reason cannot prove the demand for the sustainability of a development:

My understanding of »sustainability« is that of the sober definition of Prof. Ortwin Renn »a development in which the natural basis is maintained such that the living conditions of today's generation persist as options for the generations to come. Putting it simply: coming generations should not be much worse off than we are!

According to Renn, however, there are no scientific reasons that insist, along the lines of a quasi-automatic logic, upon the unconditional pursuance of a policy of sustainability. It is much rather an ethical decision: the necessity to select certain elements of the environment and specific life conditions that are worthy of preservation has no rationale, neither in purely economic nor in purely ecological terms. It is a matter of cultural identity: not a matter of science, but a question of ethics and politics. The selection process required for this could be made clear and interpreted by the social sciences, the accomplishment itself, however, must be rendered »ethically and politically«. More concisely: sustainability is »neither an economic nor an ecological, not even a scientific concept, but an ethical imperative.« If this is so, there is the question of the rationale for this ethical imperative. To take a stance on this is a philosophical-theological task:

Ethic is therefore more than a weighing of interests in the actual situation. Ethic aims for self-commitment, which is both absolute and generally valid. It must, however, be examined critically in the context of interest and factual constraints. For ethic cannot be without conflict. Ethical decisions are often subject to great tension, which can also stem from deep religious beliefs.

2. The limits of religion

With reference to the environmental conference, Ministerial Councilor Quennet (Federal Ministry of the Environment) already made clear in her paper in Rio (1992): consensus alone on the ecological problems, particularly among countries of the third and fourth world, cannot be achieved by appealing solely to instrumental reason, since ideological and religious factors always play a role. This also became clear at the United Nations Population Conference in Cairo (1994): The religious differences – especially on the issue of population explosion and contraception – collided fiercely here. Christian and Islamic fundamentalists did everything in their power to assert their own sexual ethics and thus maintain the status quo. Yes, Roman-Catholic fundamentalism even sought coordination with several (only a few and small) fundamentalist Muslim states (in contrast to Indonesia, Pakistan, Turkey, Egypt!), in order to continue to represent its medieval sexual ethics.

3. An ethic of responsibility instead of an ethic of success or ethic of conviction

Promoting permanent, sustainable development means, first of all, promoting the opposite of a mere ethic of success. It is thus the opposite of an action for which the end justifies the means and for which that which functions is good, bringing profit, power, enjoyment. It is precisely this that can lead to crass libertinism and Machiavellism, to wars between nations and to industrial devastation of entire landscapes. Such an »ethic« is not fit for the future.

Likewise, the demand for permanent, sustainable development can, however, also not be based on a mere ethic of conviction. Geared towards more or less isolated value ideas (justice, wealth, love, peace), a mere ethic of conviction only involves the pure inner motivation of the actor, without considering the consequences of a decision or action, the actual situation, its requirements and effects. Such an »absolute« ethic is faceless in a dangerous way; it is apolitical, but may justify even psychical or physical terrorism for reasons of conviction if need be.

The basis for the demand for sustainable development would have to be an ethic of responsibility, as proposed by the sociologist Max Weber already during the revolutionary winter of 1918/1919. Such an ethic is also not »free of conviction« with Weber, yet here there is always the question for the realistic and fore-seeable »consequences« of our action and an assumption of responsibility, thus, in principle, the inclusion of an »assessment of consequences«.

Since World War I, the knowledge and technical powers of mankind have grown immeasurably and incalculably, with very dangerous and partly irreversible long-term effects for the coming generations, as shown in the fields of nuclear energy and genetic engineering. In the late 70s, the German-American philosopher Hans Jonas therefore once again extensively thought through the »principle of responsibility« for our technological civilization, in a completely changed global situation where the survival of the human species (not Earth) is at stake. Action originating from a global responsibility for the entire biosphere, lithosphere, hydrosphere, and atmosphere of our planet! And this includes – remember the energy crisis, the depletion of nature, population growth – a self-restriction of humankind and today's freedom for the sake of survival in the future. Thus, according to Hans Jonas, a new kind of ethic is necessary to ensure the future and to show reverence for nature: an ethic of the future.

The maxim of action in view of the third millennium should thus actually be: Responsibility of the global community for its own future! Responsibility for our world and environment, but also for the world to come. Those responsible for the various world regions, and also world religions, are called upon to think in a global context and learn to act socially-minded! Indeed, this call goes in particular to the three economically leading world regions: the European Community, North America, and the Far East. They have a responsibility, which cannot be shrugged off, for sustainable development also in the other world regions: Eastern Europe, Latin America, South Asia, and, as the most highly neglected, Africa.

At the threshold to the third Millennium, the ethical cardinal question has become more urgent than ever before: Under what basic conditions can we survive, as human beings on an inhabitable Earth and shape our individual and social lives?

4. Objective and criterion: the human being in a liveable environment

A »biocentric« concept (P. W. Taylor), which ascribes a right to existence not only to individual plants and animals, but also to ecological systems and biological species, is just as inappropriate a practical aid for decisions (Birnbacher) as a »holistic concept« (K. M. Meyer-Abich), which also wants to protect inanimate nature for its own sake. Birnbacher rightly comments on the maxim: »Everyone takes consideration of everything«: »If everything merits being protected, no yardsticks are left that would justify interventions in nature«

Naturally, I do not want to represent any »anthropocentric« concept in the traditional sense, which ignores the suffering of animals and neglects the environment, but rather a humane concept in the spirit of European humanism from the Greeks to Kant, Weber and Jonas. Humanity, however, in the cosmic context, as was emphasized more in Indian and Chinese spirituality than in Christian occident since time out of mind.

Thus nothing against self-determination, self-experience, self-realization, and self-fulfillment – as long as they go hand in hand with self-responsibility and global responsibility, with the responsibility for one's fellow human beings, for society and nature, as long as they do not degenerate to narcissistic self-reflection and autistic self-centredness.

But whatever projects are planned to ensure a better future for mankind, there must be a basic ethical principle: Mankind – since Kant this formulation has been a categorical imperative – must never become a mere means to a way. It must be the ultimate end, must always remain objective and criterion. Money and capital are means to an end, as labour is a means to an end. Every technology assessment must remember: science, technology and industry are also means to an end. They too are in no way »value-free«, »neutral«, but should be assessed and used in each individual case depending on how they help humans (as individual and species) to unfold in a liveable environment.

And thereby one must remember: those who act ethically do not necessarily act uneconomically; they act – in compliance with the ethic of responsibility – in a way that prevents crisis. Several major companies had to suffer biting losses first, before they learned that the company with the longest economical success is not the company that does not care about the political and ethical implications of its production and its products, but a company that takes these aspects into account, perhaps with short-term sacrifices, and avoids severe penalties, legal restrictions, loss of public credibility, and the bad conscience of those responsible right from the start. In any case, in terms of our economy, we cannot simply rely on the market mechanisms when it comes to our long-term responsibility – as economists love to do – yet we should also not neglect the economic criterion of efficiency in case of concrete proposals.

Just as the social and ecological responsibility of the companies cannot simply be pushed onto the shoulders of politicians, moral, ethical responsibility cannot be pushed off onto religion. Ethics, which in modernity are increasingly considered to be a personal affair, must – in order to warrant the well-being of human beings and the survival of humanity – once again become a public concern of top-most significance.

216

5. No world order without a global ethic

This also applies in view of permanent, sustainable development: mankind cannot be continuously improved through ever more laws and regulations (also not by merely increasing the price of non-renewable raw materials), yet naturally also not merely through psychology and sociology. Both big and small are confronted by the same situation: Factual knowledge is no knowledge of purpose, regulation is not orientation, and laws are not yet customs. Law too needs an ethical foundation. The ethical acceptance of the law (which may be connected with sanctions by the state and implemented by force) is a prerequisite for any political culture. What use are »environmental summits« and »development summits«, new UN conventions, international agreements or even armistices, whatever, if a majority of those responsible do not even intend to abide by them, but continuously finds means and ways to represent own or collective, local, regional or national interests? The Roman adage says, »What good are laws in the absence of morals« (»Quid leges sine moribus«)?

6. The »Declaration« of the Parliament of World Religions

This text starts out from a basic insight: No new world order without a global ethic! This is a very practical answer to the question addressed by Ortwin Renn on whether we will successfully be able to »enshrine generally binding norms in our pluralist value and world«. This document bases the ethical rules on the basic directive: Every human being (whether white or coloured, man or woman, rich or poor) must be treated humanely. And the declaration reaches out beyond this seemingly self-evident demand through a second basic directive, that »golden rule« which can be found in many religious and ethical traditions of humanity for millenniums and has proved meaningful: »What you do not wish done to yourself, do not do to others!« This rule is considered the »irrevocable, unconditional norm for all areas of life, for families and communities, for races, nations, and religions.«

»Four irrevocable directives« are built on these foundations. All religions can affirm these:

1. Commitment to a culture of non-violence and respect of life (the age-old commandment: »You shall not kill« or: »Have respect for life!«).

2. Commitment to a culture of solidarity and a just economic order (the age-old commandment: »You shall not steal« or: »Deal honestly and fairly!«).

3. Commitment to a culture of tolerance and a life of truthfulness (the age-old commandment: »Your shall not lie« or: »Speak and act truthfully!«).

4. Commitment to a culture of equal rights and partnership between men and women (the age-old commandment: »You shall not commit sexual immorality« or: »Respect and love one another!«).

Passages relate directly to our problematic of a sustainable development, which concretise the first and second irrevocable directive: the »commitment to a culture of non-violence and the respect of life« as well as »commitment to a culture of solidarity and a just economic order«. Representatives of all religions affirmed the following sentences:

– »A human person is infinitely precious and must be unconditionally protected. But likewise the lives of animals and plants which inhabit this planet with us deserve protection, preservation, and care. Limitless exploitation of the natural foundations of life, ruthless destruction of the biosphere, and militarisation of the cosmos are all outrages. As human beings we have a special responsibility – especially with a view to future generations – for Earth and the cosmos, for the air, water, and soil. We are all intertwined together in this cosmos and we are all dependent on each other. Each one of us depends on the welfare of all. Therefore the dominance of humanity over nature and the cosmos must not be encouraged. Instead we must cultivate living in harmony with nature and the cosmos.«

– »If the plight of the poorest billions of humans on this planet, particularly women and children, is to be improved, the world economy must be structured more justly. Individual good deeds and assistance projects, indispensable though they be, are insufficient. The participation of all states and the authority of international organizations are needed to build just economic institutions.«

Let me finish with the closing words of this declaration, the call for a »transformation of consciousness«. Particularly in a time like ours, where the basic ethical option must apply that we commit ourselves to the well-being of coming generations and thus to sustainable development, a transformation of consciousness is in fact long overdue: »Historical experience demonstrates the following: Earth cannot be changed for the better unless we achieve a transformation in the consciousness of individuals and in public life. The possibilities for transformation have already been glimpsed in areas such as war and peace, economy, and ecology, where in recent decades fundamental changes have taken place. This transformation must also be achieved in the area of ethics and values! ... Without a willingness to take risks and a readiness to sacrifice there can be no fundamental change in our situation! Therefore we commit ourselves to a common global ethic, to better mutual understanding, as well as to socially beneficial, peace-fostering, and Earth-friendly ways of life. We invite all men and women, whether religious or not, to do the same!

VII. Closing Word

The Implementation of Sustainability brings Sustainable Prosperity

Martin Bartenstein

"Drafting the future is one of the most important and most difficult tasks the human being is faced with" argued Aurelio Peccei, the founder of the Club of Rome, in 1981. At the latest after the second oil crisis a growing number of people realised that the steady growth in economic development had come to an end.[1] Society had to accept, world wide, that new ways of securing a harmonious relationship between economic growth and the depletion of resources but also the maintenance of a social balance had become a must.

International policy as a basis for national decisions

On international level the United Nations had put "Environment" as a subject of comprehensive, global discussion for the first time on the agenda of the Stockholm Conference in 1972. But it took a further 20 years before "sustainable development" – and other topics – were directly linked to this term at the Conference in Rio de Janeiro. The participating states recognised the urgent need for action and adopted the AGENDA 21 action programme, not binding in terms of international law but containing essential corner posts for an efficient management of resources. AGENDA 21 reflects a global desire and a common political engagement for co-operation in development and environment. Responsibility for its implementation rests primarily with governments.

The euphoric spirit experienced in Rio was a short-lived though. Negotiations at the international conferences on important environment-relevant topics, organised in the course of the past decade, met with delays. Vital agreements, such as, for instance, the Kyoto Convention, were partly not ratified or not implemented. The overall situation has significantly changed since the Rio Conference in 1992. The enthusiasm for environmental concerns and the end of the cold war dominated universal thinking. The economic situation, deteriorating world wide, however, and the growing danger of terrorism created a much more pessimistic atmosphere at the Conference in Johannesburg in 2002, when a political resolution and an implementation plan were adopted but no new proposals for international environment agendas were tabled. Environmental matters lost their impetus

---

1 "Die Zukunft in unserer Hand", Aurelio Peccei, Verlag Fritz Molden Wien

in times of insecure economic climates and universal danger. Combating poverty, access to education, availability of clean drinking water became the topics on agendas. In their political resolution the heads of governments emphasised the importance of global governance – of a political world order – as a precondition for attaining the objectives of the millennium and of the AGENDA 21. Further milestones were accomplished by incorporating the results of the WTO-Conference in Doha[2] and the UN-Development Conference of Monterrey[3] into the process initiated in Rio. Contrary to events in Rio it was explicitly recognised in Monterrey that governments would not be in a position to realise the respective obligations on their own. Industry and the civil society play key roles in the field of economic growth and prosperity in order to combat poverty and environmental destruction.

The implementation plan as adopted in Johannesburg in 2002 refers above all to calling attention to the AGENDA 21 and the focal topics of combating poverty, energy, water, sustainable consumption patterns or protection of endangered species. It is not to be considered as a separate item on the list of the governments´ other obligations. In view of a targeted promotion the proposal made by OECD´s Round Table on Sustainable Development under the chairmanship of the Right Honourable Simon D. Upton, former Minister for Environment in New Zealand, is most welcome, supporting the creation of subgroups of the UN-Commission for Sustainable Development to be entrusted with the elaboration of projects for a speedy solution of problems with the required flexibility.

In addition to the world wide implementation of the local AGENDA 21 first steps at implementing the individual national sustainability strategies are being undertaken in Europe and around the world.

Implementing Rio and Johannesburg in Austria

Commissioned by Federal Chancellor Dr. Schüssel and supervised by the Federal Minister of Agriculture, Forestry, Environment, and Water Management, a national sustainability strategy was prepared in Austria on a broad discussion basis. It was subsequently adopted by the National Council in April 2002, including first implementation measures. A working programme for the implementation of the strategy was submitted for 2003 and is supported by the government.

---

2 One of the explicit objectives was a more intensive integration of development countries into the trade system.

3 0.7% of the GDP were the assessed target for national development aid.

Four fields of activities were identified in the national sustainability strategy: in addition to a good quality of living, reasonable living space and Austria's international responsibility, also a dynamic status of Austria as a business location is aimed at, in order to secure high quality economic growth, decoupled from the utilisation of resources, better jobs, social security, and a healthy and unspoilt environment for present and future generations on long-term basis. Results so far achieved in Austria are above all to be attributed to the concept of an eco-social market economy supported by all political decision carriers.

The result of this policy is reflected within the framework of a new mathematical theory as explained by Professor Radermacher. By comparing the lowest with the average income the approximate degree of social compensation can be determined by means of an equity parameter. With an equity factor of 1:1.54 (64.9%) Austria heads the parameter which means that the 30 poorest percent of Austrians get more than 20% of the total income of the country. In Brasilia, to give an example, the poorest 30 percent own a mere 9.2 percent of the national income.

In order to safeguard the high level of social peace, reflected by a significantly low strike rate, also in future, the political frame conditions will have to be adjusted to permanently changing conditions. Summarising a variety of necessary measures in all fields of politics and daily life several key areas for sustainable development are being discussed in the following.

Research and Innovation – the keys to a service society efficiently utilising resources

A new concept in national economy, tailored to the efficient utilisation of resources, as supported by Professor Weizsäcker in model factor IV, does not only use resources economically but also creates a high potential for services. New logistics systems supplementing transport logistics by the transporting and sorting of used material permit the collection of already separated used material for recycling and, in addition, create new jobs.

An excellent example for the careful handling of resources by means of new technologies is offered by the energy contracting model, a comprehensive service package for the reduction of energy input in buildings. From planning to operation of heating installations an energy (saving) contract provides for optimum service for a predetermined running period. Payment is effected with the money saved as a result of the reduction in energy consumed. The reduction of several hundred tons of CO_2-emissions per project is accompanied the savings in energy costs of, on an average, 25 to 50 %. In terms of national economy small and medium enterprises enjoy business opportunities.

Technological progress and changed demand conditions on the market are prerequisites for the survival of society. Most recent research results point towards new technologies that will essentially influence our future. Besides the growing importance of genetic research for medicinal purposes great expectations are directed at nanotechnology, beneficial applications in robotics, sensorial equipment, biotechnology, and the restoration of works of art. The Styrian Institute for Nano-structured Materials and Photonics are carrying out basic research for the custom-made manipulation and steering of light by means of nano-structured systems for increasing the efficiency of solar cells. The vast importance of technological progress is being taken into account by the Austrian government also by the respective efforts undertaken to increase research investments to 2.5% of the GNP by 2006.

Climate protection

Between 1861 and 2000 the global mean surface temperature increased by 0.6 degrees Celsius. Mean values in Austria rose by 1.8 degrees Celsius over the past 150 years.

Environment protection, by tradition, has been playing an important role in the political discourse in Austria. Austrian awareness of the importance of the environment is reflected by the fact that Austria has one of the highest share of hydropower generation in the European Union.

In order to attain the Kyoto-objective the Austrian government adopted a national climate strategy in June 2002. Austria was allocated a reduction of greenhouse emissions by 13% until 2012 in accordance with the respective EU-agreement. With a view to reaching this ambitious target the required measures as regards implementation and reform have been adopted primarily by

- rerouting heavy traffic from road to rail (introduction of the km-dependent lorry toll, realisation of important rail infrastructure projects),
- the extension of environment protection funds to climate-relevant investment and programmes (above all renewable energy carriers and energy efficiency),
- tailoring tax systems to ecology, and
- increasing the share of renewable energy carriers in electricity generation from 70 to 78%. In the EU-comparison Austria´s share in renewable energy ranks first already.

224

An essential factor for this development has been the promotion of renewable energy for the generation of electricity by the eco-power act adopted in 2002, targeting at 78.1% coverage for Austria as laid down in the EU-Directive on the promotion of renewable energy carriers.

Within the framework of the 2003 budget accompanying law the Austrian JI/CDM-programme[4] was initiated. It offers enterprises the realisation of greenhouse emission reduction projects abroad by means of concluding contracts for the purchase of the reduction units achieved by the government – an entirely novel form of project financing, opening up new chances in the field of environment and energy technologies on the international market. At the same time national costs for attaining the Kyoto-objective can be reduced with the help of private industry.

The implementation of the national climate strategy furthermore looks promising as regards its effects on the national economy. According to a study undertaken by WIFO (Austrian Institute for Economic Research)[5] an increase of the order of 20.000 to 26.000 jobs can be expected in energy carriers, building restoration, and public commuter transport for the period 2005 to 2010.

In the wake of the eco-power act the so-called eco-power ordinances were decreed and have become effective as at 1/1/2003. They contain fixed prices for electrical energy from eco-power plants on the one hand, to promotional payments as well as a power-heat coupling surcharge to be levied for the supply of currency to the end consumer, on the other.

Policies implemented in agriculture and forestry

11% of agriculturally useful areas are utilised according to biological criteria in Austria, about every 10th farmer being a bio-farmer. In order to further support the trend towards biological farming a bio-action programme has been concluded and is gradually being implemented. Approximately € 7.5 m more were paid for bio-promotions out of the agrarian environment programme (ÖPUL 2000) in 2002 than it had been conceded in 2001. The bio-action programme elaborated on EU-level was significantly supported by Austria.

In 2002 promotion granted for biomass and renewable energy carriers was raised by altogether € 15m, permitting the promotion of biomass heating plants at a scale never experienced before. The impetus of this promotion will not only be reflected in improved climate protection but will also bring about an economic

---

4 Joint Implementation/Clean Development Mechanism
5 "Energieszenarien bis 2010"

boom. Investments of approx. € 50m can be expected, creating about 1.000 additional jobs in the construction industry and new long-term employment in rural areas. Farmers will moreover be given additional sales opportunities for biomass products.

Social environment aspects

Securing participation in active employment has become a great concern for people above the age of 60 in Austria too. Getting older should not be seen as an imminent danger to one's existence. In the light of the demographic development – as shown by Professor Schmid – permanent adjustments are necessary in the field of health and social security, above all as regards old age provisions. This philosophy has been incorporated as a long-term objective for the promotion of development possibilities for all generations in the national sustainability strategy.

Forecasts of future expenditure on old age provision show alarming increases to the amount of 3 to 5 percent points of the GNP in most countries (4 1/2% for Austria), in some countries even more. Increases of such order give priority to the question of the best possible financing modus of pension systems. In conjunction with the key points referred to by Professor Rürup measures will have to be adopted in the following areas:

Streamlined policies in economy, employment, education, and family affairs

Since any financing of old age provision depends primarily on the overall economic development (growth, labour productivity, employment), an active employment and labour market policy, by which unemployment is abolished and employment levels are raised, is gaining ever more importance as regards the financing of old age pensions.

The political challenge lies in steering the respective frame conditions – under the headings of "Coordinating job and family", "Company-internal further training of older employees", etc. The present programme of the Austrian government contains a number of respective measures and several additional ones are in preparation. To quote an example: non-wage costs are to be reduced by 3% for employees above 56/58 years of age, and by 10% for those above 60. Under certain conditions, e.g. in the case of a job at risk, the possibilities of the active labour market policy for qualifying older employees shall be comprehensively employed. Unemployed persons under 25 or older than 50 shall be offered a reasonable job within a period of eight weeks. Where this proves impossible the unemployed person shall be entitled to participate in a qualification measure.

Education

The internationally acknowledged and highly differentiated Austrian education system can be mentioned as an explanation for the fact that Austria has had the lowest youth unemployment rate in the EU since the beginning of 2003.

One of the national particularities of the country's educational pattern is the "dual education system" in which young people receive a combination of theoretical and practical instructions. They get training in an enterprise while simultaneously attending classes at a technical college for, on an average, three years. In order to further expand this successful system, which offers an excellent potential of qualified workers, Austria has created new instruction courses (personnel services, crystal cutting technology, sanitation and air-conditioning technology, eco-energy installations, mechatronics, etc.).

Austria's top position in the EU-comparison is also to be attributed to intermediate and secondary professional schools in the fields of fashion, agriculture and forestry, tourism, and commerce. At intermediate professional schools general education plus professional qualifications are obtained within three to four years. A leaving certificate from a secondary professional school (five years) permits access to further studies at a university in addition to having obtained a professional qualification. The "training firm" as instruction place and additional instruction method, obligatory for pupils at trade schools and colleges since 1993, is becoming ever more popular in other school too. Some 1.200 training firms are operative in Austria at the time being.

Universities of Technology and trade-oriented Technical Institutes offer projects in the field of engineering and technology, where professional qualification is combined with important economic key competences. Diploma theses accomplished under the supervision of experienced tutors in co-operation with economy and industry have proved most successful in the past years.

Besides studying at a university there is now also the possibility of attending a professional quasi-university. The novelty in this respect is a university curriculum plus a professional practice in, for instance, tourism, informatics, or information technology, media, design, health, or social affairs. They are administrated by legal persons of private or public law upon approval by the council for advanced studies. Altogether 7,497 students have graduated from such institutes since 1993.[6]

Future aspects in general

The results so far achieved in industrialized countries are conspicuous. But the challenges in store for us world wide are enormous and will only by manageable by internationally concerted efforts.

The generally weak economic growth, increasing population figures, climate changes, and security concerns should not be considered separately. The end of steady and massive economic growth commenced in the 70ies of the 20th century as Professor Giarini summarised in his chapter. In terms of national economy future generations will require primarily services, traditional industrial sectors are on the decline.

One of the essential basic conditions for a sustainable global future is fair global competition. It should therefore be the explicit objective of national governments to support the frame conditions for the WTO-rules and in the UN-agreements.

A new multilateral trade round was initiated at the ministerial conference in Doha in 2001. Not only new areas such as trade and environment or transparency in public purchasing are to be incorporated into the WTO-rules but also the demands and requirements of developing countries. At the ministerial meeting in Cancun (Mexico) in September 2003 important decisions would have been to be anticipated. The negotiation's stop in Cancun is pitiable and a setback to the efforts of the WTO, namely for developing countries. The gap between the industrialized and developing countries was close as never before, especially in the field of agriculture.

Sustainability is a central political challenge of the 21 st century. Economic policies will have to play a particularly important role as regards a comprehensively understood sustainability. The human being must be the centre in all relevant areas, be it trade, research, or innovation. As a consumer each of us is entitled to environmentally sound products and production systems, due attention being paid to the efficient use of resources. Producers are responsible for the ways in which this is done. The relationship between economic and environmental policies, often regarded as tense, could be appeased in a creative way with the sustainability concept because of the common objectives involved. In this

---

6 2003 Economic Report for Austria, published by the Federal Ministry of Economics and Labour

way the objectives of a "sound environment" and "economic prosperity" will become complementary objectives of economic politics.

References

Neoliberalism against Sustainable Development? (Franz Josef Radermacher)

Affemann, N., B.F. Pelz und F.J. Radermacher: Globale Herausforderungen und Bevölkerungsentwicklung: Die Menschheit ist bedroht. Beitrag für den Beirat der Deutschen Stiftung Weltbevölkerung e. V., Landesstelle Baden-Württemberg, 1997

Brown, G.: Tackling Poverty: A Global New Deal. A Modern Marshall Plan for The Developing World. Pamphlet based on the speeches to the New York Federal Reserve, 16 November 2001, and the Press Club, Washington D.C., 17 December 2001. HM Treasury, February 2002

Club of Rome (ed.): No Limits to Knowledge, but Limits to Poverty: Towards a Sustainable Knowledge Society. Statement of the Club of Rome to the World Summit on Sustainable Development (WSSD), 2002

Gore, A.: Wege zum Gleichgewicht – Ein Marshallplan für die Erde. S. Fischer Verlag GmbH, Frannkfurt, 1992

Information Society Forum (ed.): The European Way for the Information Society. European Commission, Brussels, 2000

Kämpke, T., F.J. Radermacher, R. Pestel: A computational concept for normative equity. European J. of Law and Economics 15, 129-163, 2002

Küng, H.: Projekt Weltethos, 2. Aufl., Piper, 1993

Küng, H. (ed.): Globale Unternehmen – globales Ethos. Frankfurter Allgemeine Buch, Frankfurt, 2001

Neirynck, J.: Der göttliche Ingenieur. expert-Verlag, Renningen, 1994

Radermacher, F.J.: Globalisierung und Informationstechnologie. In: Weltinnenpolitik. Intern. Tagung anläßlich des 85. Geburtstages von Carl-Friedrich von Weizsäcker, Evangelische Akademie Tutzing, 1997. In (U. Bartosch und J. Wagner, eds.) S. 105-117, LIT Verlag, Münster, 1998

Radermacher, F.J.: Die neue Zukunftsformel. bild der wissenschaft 4, S. 78ff., 2002

Radermacher, F.J.: Balance oder Zerstörung: Ökosoziale Marktwirtschaft als Schlüssel zu einer weltweiten nachhaltigen Entwicklung. Ökosoziales Forum Europa (ed.), Wien, August 2002, ISBN: 3-7040-1950-X

Schauer, T. F.J. Radermacher (eds.): The Challenge of the Digital Divide: Promoting a Global Society Dialogue. Universitäts-Verlag, Ulm, 2001

Schmidt, H.(Hg.): Allgemeine Erklärung der Menschenpflichten – Ein Vorschlag. Piper Verlag GmbH, München, 1997

Schmidt, H.: Die Selbstbehauptung Europas. Perspektiven für das 21. Jahrhundert. Deutsche Verlags-Anstalt, Stuttgart, 2000

Schmidt-Bleek, F.: Wieviel Umwelt braucht der Mensch? MIPS – Das Maß für ökologisches Wirtschaften, Birkhäuser Verlag, 1993

Töpfer, K.: Kapitalismus und ökologisch vertretbares Wachstum – Chancen und Risiken. in: Kapitalismus im 21. Jahrhundert, S. 175-185, 1999

Töpfer, K.: Ökologische Krisen und politische Konflikte. in: Krisen, Kriege, Konflikte (A. Volle und W. Weidenfeld, ed.), Bonn, 1999

Töpfer, K.: Environmental Security, Stable Social Order, and Culture. in: Environmental Change and Security Project Report, Woodrow Wilson Centre, No. 6, 2000

Töpfer, K.: Globale Umweltpolitik im 21. Jahrhundert, eine Herausforderung für die Vereinten Nationen. in: Erfurter Dialog (Thüringer Staatskanzlei, ed.), 2001

von Weizsäcker, C. F.: Bedingungen des Friedens. Vandenhoeck und Ruprecht, Göttingen, 1964

von Weizsäcker, E.U., A. B. Lovins, L. H. Lovins: Faktor Vier: doppelter Wohlstand, halbierter Naturverbrauch. Droemer-Knaur, 1995

Good Global Governance (Petra C. Gruber)

Altvater, Elmar, Mahnkopf, Birgit: Grenzen der Globalisierung; Ökonomie, Ökologie und Politik in der Weltgesellschaft. Münster 1996.

Amon, Werner, Liebmann, Andreas (Hrsg.): Umwelt – Friede – Entwicklung. Dimensionen 2000, Wien 1997.

Beck, Ulrich: Perspektiven der Weltgesellschaft. Frankfurt am Main 1998.

Bourdieu, Pierre: Gegenfeuer; Wortmeldungen im Dienste des Widerstands gegen die neoliberale Invasion. Konstanz 1998.

Esteva, Gustavo: FIESTA – jenseits von Entwicklung, Hilfe und Politik. Frankfurt a. M. / Wien 1992.

Grober, Ulrich: Konstruktives braucht Zeit. Aus Politik und Zeitgeschichte, B31-32/2002.

Kabou, Axelle: Weder arm noch ohnmächtig; Eine Streitschrift gegen schwarze Eliten und weiße Helfer. Basel 1995.

Kennedy, Margit: Gefahren und Chancen der Globalisierung. In: Zeitschrift für Sozialökonomie 121/1999. S.27 ff.

Madörin, Mascha: Männliche Ökonomie – Ökonomie der Männlichkeit. In: Bundesministerium für Frauenangelegenheiten, Frauenwirtschaftskonferenz, Schriftenreihe BD.6, Wien 1995.

Mascha, Andreas: Das Holistische Marketing – Eine zukunftsfähige Managementstrategie. Sonderschrift der Arbeitsgruppe Integrales Management / Institut homo integralis.

Messner, Dirk, Nuscheler, Franz: Das Konzept Global Governance Stand und Perspektiven; INEF Report, Duisburg, Heft 67/2003.

Moser, Anton, Riegler, Josef: Konfrontation oder Versöhnung? Ökosoziale Politik mit der Weisheit der Natur. Graz / Stuttgart 2001.

Narr, W., D., Schubert, A.: Weltökonomie, Die Misere der Politik. Frankfurt am Main 1994.

Nohlen, D., Nuscheler, F. (Hrsg.): Handbuch der Dritten Welt; 1 Grundprobleme, Theorien, Strategien. Bonn 1993.

Nuscheler, Franz (Hrsg.): Entwicklung und Frieden im Zeichen der Globalisierung, Bonn 2000.

Sachs, Wolfgang (Hg.): Wie im Westen so auf Erden, Ein polemisches Handbuch zur Entwicklungspolitik. Reinbek bei Hamburg 1993.

Schmee, J., Weissel, E. (Hrsg.): Die Armut des Habens; Wider den feigen Rückzug vor dem Neoliberalismus. Wien 1999.

Stiftung Entwicklung und Frieden (SEF) (Hrsg.): Nachbarn in Einer Welt; Der Bericht der Kommission für Weltordnungspolitik. Bonn 1995.

Woyke, Wichard (Hrsg.): Handwörterbuch Internationale Politik. Opladen 1993.

The Limits of Monetisation – Changes in the Money-Driven Society (Patrick M. Liedtke)

Beck, U. (1997): Erwerbsarbeit durch Bürgerarbeit ergänzen, in: Kommission für Zukunftsfragen der Freistaaten Bayern und Sachsen, Band III, S. 146-168.

Beck, U. (1999): Schöne neue Arbeitswelt. Vision: Weltbürgergesellschaft, Frankfurt a.M./New York.

Britton, F. (1994): Rethinking Work – An Exploratory Investigation of New Concepts of Work in a Knowledge Society. Paris.

Competitiveness Advisory Group (1995 and 1996): First to Third Report to the President of the European Commission, the Prime Ministers and Heads of State. Brussels.

Delsen, L./ Reday-Mulvey, G. (1996): Gradual Retirement in the OECD Countries.

Dollase, R. et.al. (1999): Zeitstrukturierung unter hypothetischen Bedingungen der völligen Wahlfreiheit oder: Das Flexibilisierungsparadoxon.

Employee Benefit Research Institue (1996-2001): Monthly Newsletters. New York.

European Commission (1993): Actions for Stimulation of Transborder Telework & Research Cooperations in Europe. Brussels.

European Commission (1994-2000): Employment in Europe. Brussels. And follow-up documentation.

Giarini, O. & Liedtke, P. (1997 et. al.): Wie wir arbeiten werden – ein Bericht an den Club of Rome. Hamburg, Bilbao, Paris, Rom, et. al.

Gruhler, W. (1990): Dienstleistungsbestimmter Strukturwandel in deutschen Industrieunternehmen. Köln.

ILO (1994-2000): World Labour Reports. World Employment Reports. Yearbooks of Labour Statistics. Geneva.

International Social Security Association (various): International Social Security Review. Various Issues. Geneva.

Liedtke, P. (2001): "Erwartungen der Erwerbstätigen an die Arbeitsbedingungen des 21. Jahrhunderts", in: Bensel, N.: „Von der Industrie- zur Dienstleistungsgesellschaft", Frankfurt.

Liedtke, P. (2001): "The Future of Active Global Ageing: Challenges and Responses", in Geneva Papers on Risk and Insurance – Issues and Practice, vol. 26, no. 3, July 2001.

OECD (1994): Labour Force Statistics 1972-1992. Paris

OECD (1999): Implementing the OECD Jobs Strategy – Assessing Performance and Policy. Paris.

OECD (various): Employment Outlook. Paris.

Rauschenbach, Th. (1999): Rede anlässlich der Mitgliederversammlung des Deutschen Vereins für öffentliche und private Fürsorge e.V. am 1. Dezember 1999 in Frankfurt/Main.

S. Simon, G. (1999): Zeit-Geist-Wende. In: Kommune, Nr. 8, August 1999.

Scherrer, K. und Wieland, R. (1999): Belastungen und Beanspruchung bei der Arbeit im Call Center. In: Gesina aktuell, Nr. 2, April 1999.

UNDP (1994-2000): Human Development Reports. New York.

Wieland, R. (1999): Arbeitswelt 2000 – Kreativ, motiviert, flexibel.

Risk and Sustainability (Walter R. Stahel)

Amalberti, R. (1994) Quand l'homme et la machine ne se comprennent plus. In: Bulletin deliaison de l'Institut Fredrik R. Bull, Louveciennes.

Berliner, B. (1982) Limits of insurability of Risks, Zurich, Swiss Reinsurance Company, Prentice-Hill Inc.,Englewood Cliffs,N.J.

Bernold, T. (1990) Industrial Risk Management: A Life-cycle Engineering Approach, Proceedings of a Conference at the Swiss Federal Institute of Technology, Zurich, In: Elsevier Amsterdam a Journal of Occupational Accidents, Vol.13 (1/2)

Carnoules Declaration of the Factor 10 Club (1994).

Dieren, Wouter van (1995) Taking Nature into account, Birkhäuser-Verlag, Basel, ISBN 3-7643-5173-X

ETAN Expert Working Group for the European Commission Directorate General XII, Environment and Climate RTD Programme, Targeted Socio-Economic Research Programme, ETAN Working Paper on Climate Change and the Challenge for Research and Technological Development (RTD) Policy. Prepared by an independent Final Report – December 1998

Ewald, F. (1989) Die Versicherungs-Gesellschaft, In: Kritische Justiz, 22 Jahrgang, Heft 4,Baden-Baden

Giarini, Orio und Stahel, Walter R. (1989/1993) The Limits to Certainty, facing risks in the new Service Economy, 2nd edn; Kluwer Academic Publishers, Dordrecht, Boston, London – ISBN 0-7923-2167-7. (2002) Die Performance Gesellschaft: Chancen und Risiken beim Übergang zur Service Economy. Metropolis Verlag Marburg. ISBN 3-89518-320-2

Gross (1995), Die Multi-Options-Gesellschaft, Suhrkamp Verlag

Gruhler, Wolfram (1990) Dienstleistungsbestimmter Strukturwandel in deutschen Industrieunternehmen; Deutscher Instituts Verlag Köln; ISBN 3-602-24406-7.

Haller, M. and Petin, J. (1994) Geschäft mit dem Risiko – Brüche und Umbrüche in der Industrieversicherung; in: Schwebler et al (Hrsg.) Dieter Farny und die Versicherungswirtschaft, Verlag Versicherungswirtschaft Karlsruhe

Nutter, F.W., (1994) The Role of Government in the United States in Addressing Natural Catastrophes and Environmental Exposure, In: Geneva Papers, 19th year, No. 72, July, p.244

Schmidt-Bleek, Friedrich (1996) The Fossil Makers – Factor 10 and more. Birkhäuser Verlags-AG, Berlin, Basel; ISBN 3-7643-2959-9

Schmid, G. (1990)Rechtsfragen bei Großrisiken, In Zeitschrift für Schweizerisches Recht, NF 109 (1990) sowie in den Proceedings des Schweizerischen Juristentages

Showater, Sands, P. and Myers, M.F. (1994) Natural Disasters in the United States as Release Agents of Oil, Chemicals or Radiological Materials between 1980-1989, Analysis and Recommendations, In Risk Analysis, Vol. 14, No.2, p.169ff

Sogh, G. and Fauve, M. (1991) Compensation for Damages Caused by Nuclear Accidents: A Convention as Insurance, Etudes et Dossiers No. 156, July 1991, Geneve, The Geneva Association

Stahel, Walter and Reday, Geneviève (1976/1981) Jobs for Tomorrow, the potential for substituting manpower for energy; report to the Commission of the European Communities, Brussels/Vantage Press, N.Y.

Stahel, Walter R. (2000) Incentives for loss prevention instead of disaster management by the State in case of catastrophic risks; in: Coles, E., Smith D. and Tombs S. (eds.) Risk Management and Society, pp. 81-100 (1995) 300 examples of higher resource productivity in today's industry and society (Intelligente Produktionsweisen und Nutzungskonzepte) – Handbuch Abfall 1 – Allg. Kreislauf- und Rückstandswirtschaft; Band 1 und 2, Landesanstalt für Umweltschutz Baden-Württemberg (Hrsg.), Karlsruhe

(1994) The impact of shortening (or lengthening) of life-time of products and production equipment on industrial competitiveness, sustainability and employment; research report to the European Commission, DG III, November 1994

(1985) Hidden innovation, R&D in a sustainable society, in: Science & Public Policy, Journal of the International Science Policy Foundation, London; Volume 13, Number, 4 August 1986: Special Issue : The Hidden Wealth.

(1984) "The Product-Life Factor"; in: Orr, Susan Grinton (ed.) An Inquiry into the Nature of Sustainable Societies: The Role of the Private Sector; HARC, The Woodlands, TX

World Health Organization, (1991) Report on Chernobyl: see Risk Management Newsletter No.10 (January 1991), Geneva, The Geneva Association.

The Generation-Contract (Bernd Rürup und Jochen Jagob)

Aaron, Henry, 1966, The Social Insurance Paradox, Canadian Journal of Economics and Political Science, 32, 371-374

Auerbach, Alan, Laurence Kotlikoff, Willi Leibfritz, 1999, Generational Accounting Around the World, National Bureau of Economic Research Project, The University of Chicago Press

Barr, Nicholas, 2000, Reforming Pensions: Myths, Truths, and Policy Choices, IMF Working Paper WP/00/139

Breyer, Friedrich, 2001, Why funding is not a solution to the "Social Security Crisis", DIW Discussion Paper No. 254

Breyer, Friedrich, Mathias Kifmann, 2001, Incentives to retire later – a solution to the social security crisis? DIW Discussion Paper No. 266

Breyer, *Friedrich*, 1989, On the Intergenerational Pareto Efficiency of Pay-as-you-go Financed Pension Systems, Journal of Institutional and theoretical Economics, Journal of Institutional and Theoretical Economics, 145, 643-658

Brunner, Johann, 1996, Transition from a pay-as-you-go to a fully funded pension system: The case of differing individuals and intragenerational fairness, Journal of Public Economics, 60, 131-146

Brunner, Johann, 1994, Redistribution and the Efficiency of the Pay-as-you-go Pension System, Journal of Institutional and theoretical Economics, 150, 511-523

Diamond, Peter, 1977, A Framework for social security analysis, Journal of Public Economics, 8, 275-298

European Commission, 2001, Reforms of Pension Systems in the EU – An Analysis of the Policy Options -, European Economy, No. 73,171-222

European Commission, 1999, Generational Accounting in Europe, European Economy, No. 6

Feldstein, Martin, 2001, The Future of Social Security Pensions in Europe, NBER Working Paper No. 8487

Feldstein, Martin, 1996, The Missing Piece in Policy Analysis: Social Security Reform, American Economic Review, 86, 1-14

Fenge, Robert, 1995, Pareto-efficiency of the Pay-as-you-go Pension System with Intragenerational Fairness, Finanzarchiv, NF 52, 357-363

Herbertsson, Tryggvi Thor, J. Micheal Orszag, 2001, The costs of Early Retirement in the OECD, Institute of Economic Studies, University of Iceland, Working Paper W01:02

Lindbeck, Assar, Mats Persson, 2002, The Gains from Pension Reform, Institute for International Economic Studies, Seminar Paper No. 712

Merton, Robert, 1983, On the Role of Social Security as a means for efficient Risk Sharing in an Economy where Human Capital is not tradeable, in Z. Bodie, J. B. Shoven (eds.): Financial Aspects of the United States Pension System, 325-358, University of Chicago Press

Orszag, Peter, Joseph Stieglitz, 1999, Rethinking Pension Reform: Ten Myths about Social Security Systems, Paper Presented at the conference on "New Ideas about Old Age Security", The World Bank, Washington DC

Persson, Mats, 2000, Five Fallacies in the social security debate, Institute for International Economic Studies, Stockholm University, Seminar Paper No. 686

Riphahn, Regina, Peter Schmidt, 1999, Lockt der Ruhestand oder drängt der Arbeitsmarkt? Langfristige Entwicklung der Gesetzlichen Rentenversicherung und Determinanten des Rentenzugangs, in E. Wille (Hrsg.): Entwicklung und Perspektiven der Sozialversicherung, p. 101-145, Nomos Verlag, Baden Baden

237

Rürup, Bert, Patrick M. Liedke, 1998, Umlageverfahren versus Kapital-deckungsverfahren, in Cramer, Förster, Ruland (Hrsg.): Handbuch der Alters-versorgung, S.779-798, Fritz Knapp Verlag, Frankfurt am Main

Samuelson, Paul, 1958, An Exact Consumption Loan Model of Interest with or without Social Contrivance of Money, Journal of Political Economy, 66, 467-482

Siddiqui, Sinkandar, 1997, The Pension Incentive to retire: Empirical Evidence for West Germany, Journal of Population Economics, 10, 337-360

Sinn, Hans-Werner, 1999a, The Crisis of Germany´s Pension Insurance system How it can be resolved, CESifo Working Paper Series No. 191

Sinn, Hans-Werner, 1999b, Pension Reform and Demographic Crisis: Why a funded System is needed and why it is not needed, Paper presented at the 55th IIPF Congress in Moscow

Thum, Marcel, Jakob von Weizsäcker, 2000, Implizite Einkommensteuer als Messlatte für aktuelle Rentenreformvorschläge, Perspektiven der Wirtschaftspolitik, 1, 453-468

Worldbank, 1994, Averting the old age crisis, Oxford University Press

Society and Ethos (Hans Küng)

Die Grundlage für die folgenden grundsätzlichen Ausführungen bildet mein Buch »Projekt Weltethos« (München 1990)

O. Renn, Ökologisch denken – sozial handeln: Zur Realisierbarkeit einer nach-haltigen Entwicklung (Redemanuskript 1994)

D. Birnbacher – C. Schicha, Vorsorge statt Nachhaltigkeit – ethische Grundlagen der Zukunftsverantwortung (Redemanuskript 1994)

Dietmar Mieth, Aufsatz: Theologisch-ethische Ansätze im Hinblick auf die Bioethik, in: Concilium 25 (1989) Heft 3 (als ganzes der Thematik »Ethik der Naturwissenschaften« gewidmet)

H. Jonas, Das Prinzip Verantwortung. Versuch einer Ethik für die technologische Zivilisation, Frankfurt 1984,

Der Spiegel, Nr. 9, 1993

H. Küng – K.-J. Kuschel (Hrsg.), Erklärung zum Weltethos. Die Deklaration des Parlamentes der Weltreligionen, München 1993, Kap. II: Grundforderung: Jeder Mensch muss menschlich behandelt werden, vgl. Erklärung zum Weltethos, Kap. III: Vier unverrückbare Weisungen, vgl. Erklärung zum Weltethos, Kap. III,1: Verpflichtung auf eine Kultur der Gewaltlosigkeit und der Ehrfurcht vor allem Leben, vgl. Erklärung zum Weltethos, Kap. III,2:

Verpflichtung auf eine Kultur der Solidarität und eine gerechte Wirtschafts-
ordnung, vgl. Erklärung zum Weltethos, Kap. IV: Wandel des Bewusstseins.
H. Küng, Das Christentum, 1992 (Neuauflage) Piper Verlag, München

Peter Lang · Europäischer Verlag der Wissenschaften

Walter Leal Filho (ed.)

International Experiences on Sustainability

Frankfurt am Main, Berlin, Bern, Bruxelles, New York, Oxford, Wien, 2002.
238 pp., num. fig., num. tab.
Environmental Education, Communication and Sustainability.
Edited by Walter Leal Filho. Vol. 12
ISBN 3-631-50110-2 / US-ISBN 0-8204-6045-1 · pb. € 34.80*

There is a wide range of projects, institutions and initiatives in the field of sustainable development which have been taking place and which have provided a concrete contribution to the cause of sustainability as a whole. Unfortunately, most are little known. This perceived need for information on international perspectives on sustainability is addressed in this book. It gathers a set of papers which provide a synthetic overview of the effectiveness of implementation activities and initiatives at various levels, bringing together various clusters of organisations and a wide range of approaches. Whilst a special emphasis is given to Germany, from where reports on the work of the German Council for Sustainable Development, the German International Co-operation Agency and the German Federal Environment Foundation are documented, examples of other initiatives taking place elsewhere in Europe and North America are provided. Contributors to this book have made a great deal of efforts in providing synthetic overviews of the effectiveness of implementation activities at different levels, drawing lessons with a wide geographical scope, broad relevance and wider implications and applications. Readers will notice that some major new developments and outstanding problems needing further attention are outlined. Last but not least, the book provides concrete examples of governmental, inter-governmental, and non-governmental successful or promising efforts, showing that sustainability as both a process and as a goal may be pursued in different ways.

Contents: Sustainability: world trends and future perspectives · Sustainable Development as a Framework for Technical Co-operation · Participation

Frankfurt am Main · Berlin · Bern · Bruxelles · New York · Oxford · Wien
Distribution: Verlag Peter Lang AG
Moosstr. 1, CH-2542 Pieterlen
Telefax 00 41 (0) 32 / 376 17 27

*The €-price includes German tax rate
Prices are subject to change without notice
Homepage http://www.peterlang.de